Simple Physics Experiments with Everyday Materials

Judy Breckenridge

Illustrated by
Frances Zweifel

Sterling Publishing Co., Inc. New York

Acknowledgments

I would like to thank Martha Register of the Seminole County High School Physics Department in Donalsonville, Georgia, and Dwight Myers of the Greeneville High School Science Department in Greeneville, Tennessee, for their help with some of the experiments.

Also, a special "thank you" to my editor, Claire Bazinet, for her suggestions, and to my parents, Kathryn and Bob Watts, for their encouragement.

Library of Congress Cataloging-in-Publication Data

Breckenridge, Judy.
Simple physics experiments with everyday materials / Judy Breckenridge ; illustrated by Frances Zweifel.
p. cm.
Includes index.
Summary: Includes over sixty simple experiments which provide information about heat, air, light, sound, gravity, and more.
ISBN 0-8069-8606-9
1. Physics—Experiments—Juvenile literature. 2. Scientific recreations—Juvenile literature. [1. Physics—Experiments. 2. Experiments. 3. Scientific recreations.] I. Zweifel, Frances W., ill. II. Title.
QC25.B74 1993
530′.078—dc20 92-25312
CIP
AC

10 9 8 7 6 5 4 3 2 1

Published in 1993 by Sterling Publishing Company, Inc.
387 Park Avenue South, New York, NY 10016

Distributed in Canada by Sterling Publishing
c/o Canadian Manda Group, P.O. Box 920, Station U
Toronto, Ontario, Canada M8Z 5P9
Distributed in Great Britain and Europe by Cassell PLC
Villiers House, 41/47 Strand, London WC2N 5JE, England
Distributed in Australia by Capricorn Link Ltd.
P.O. Box 665, Lane Cove, NSW 2066
Manufactured in the United States of America

Sterling ISBN 0-8069-8606-9 Trade
0-8069-8607-7 Paper

To my husband,
Rufus G. Breckenridge, M.D.,
for his patience, understanding, and
knowledge of physics,
and to my daughters,
Suzanne, Audra, Caroline, and Jena,
who "tried out" many of the experiments
in this book

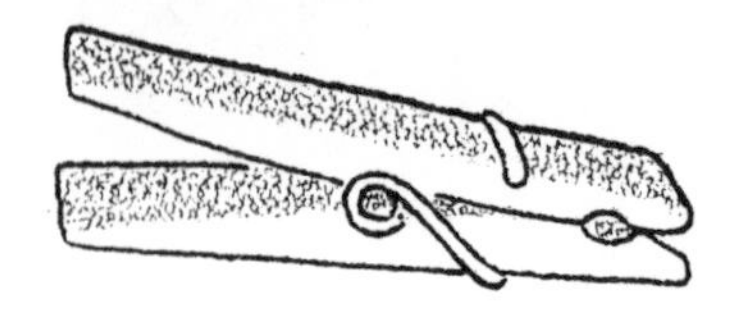

Contents

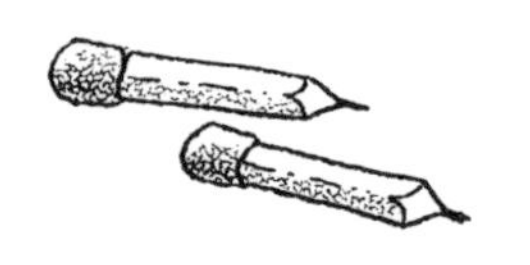

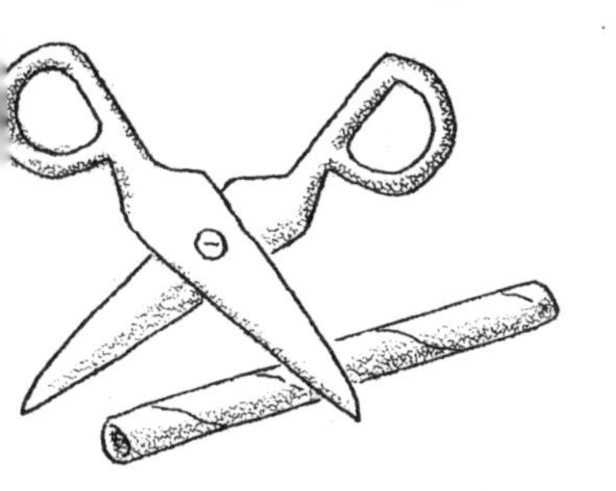

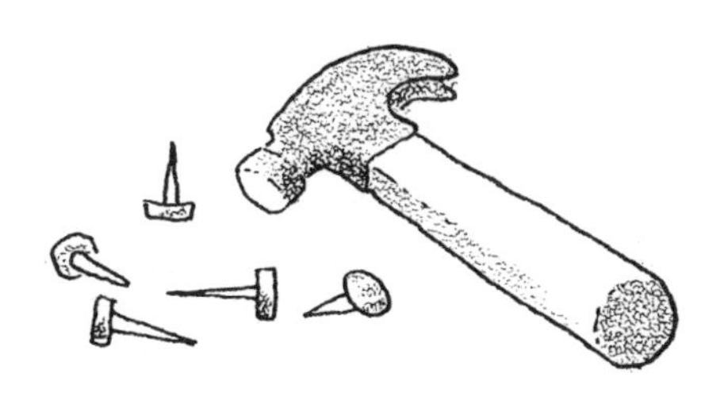

Before You Begin

It's a marvelous world that we live in. A world filled with rainbows and rockets, with echoes and electric sparks, with atomic particles and planets, and with invisible forces and vibrations that affect you without your even knowing they exist.

Physics comes from the Greek word *physica*, meaning "natural things." Learning about the natural things of this world, what is happening every second all around us, is what physics is all about. What better way to learn about physics than in our everyday laboratory—the world itself—where it can be experienced and not just studied.

These hands-on experiments put the natural world at your fingertips. And you don't have to spend time and money at the story buying expensive equipment and supplies. You can do all of these experiments using just odds and ends that you probably have around your home. (If you don't have an eyedropper, use a straw. See instructions on page 63.) Too, unless it is absolutely necessary so that you can do the experiment correctly, we don't give you exact measurements, like length of string or size of can to use. Use whatever you have available and, if necessary, adjust or substitute using the experiment's instructions as a guide. Most likely the experiment will work just as well. And, if it doesn't, with some thought you'll discover *why* it doesn't and be well on your way to becoming a problemsolver (someone who gets things done no matter what).

If you do have trouble getting an experiment to work right away, don't give up. Re-read the instructions, give it another try, and you'll probably succeed. We did, or the experiment wouldn't be in this book. Even if you can't get the experiment to do exactly what we say happens, maybe while you are trying you will change the experiment and do something even more fantastic, that we never even thought of. The possibilities are endless. Every answer leads to another question, that cries out for another answer. Just think of the fun you can have learning, like students of long-ago Socrates, without lectures, worksheets, or tests.

Several of these experiments require the use of heat. Where possible, the heat needed comes from a 100-watt light bulb with a foil shade (the shade makes it easier on your eyes and the light bulb is safer to use). A symbol reminding you to be careful marks those experiments using heat. Instructions for making the foil shade are on page 8.

You will also find that some of these experiments are more serious and cover important scientific principles, while others are tricks or fun activities with just a little basic physics thrown in. All are designed, however, to help you learn and provide hours of enjoyment, so turn the page and search out what you need.

Happy experimenting!

NOTE: This symbol will remind you to be extra careful at certain times when you will be using heat in these experiments. Always remember to use pliers, tongs, an oven mitt, or a pot holder to handle anything that may be hot. Even minor burns can be painful.

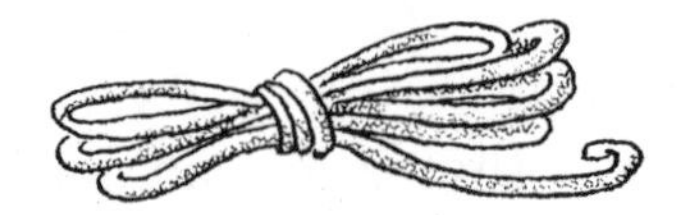

MAKE A FOIL LAMP SHADE

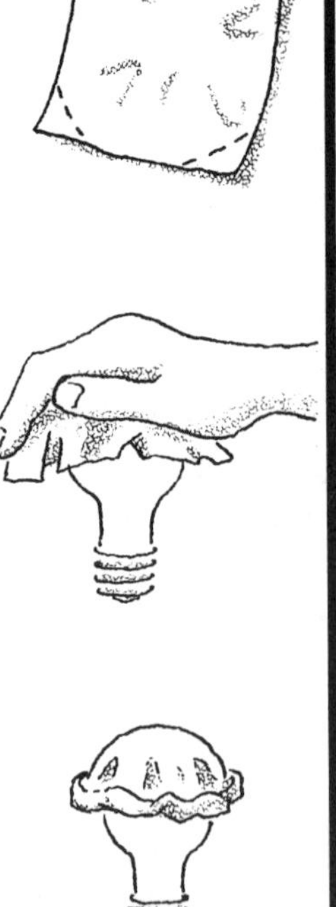

Some of the experiments in this book use a lamp with a 100-watt bulb as a source of heat. To collect and direct the heat from the bulb towards the experiment, make a cap-shaped lamp shade out of aluminum foil.

Take a small square of foil and round off the corners by folding or cutting. Place the foil on top of the *cold* (not lit) light bulb and shape it into a cap shape. Turn the edges of the foil up slightly, and you have a shade ready for your experiments with heat. Remember to watch for the HOT symbol (see above), and be careful.

BEATING THE HEAT

Heat is a form of energy. It is what happens when molecules move around within a substance. The faster the molecules move, the hotter the substance gets.

All things have some heat. Heat passes freely from hot to cold things. For example, you pour a hot drink into a glass with ice cubes. The heat from the drink moves into the ice, melting it, and the drink is cooled, meaning that some of the heat has been removed.

Heat affects what kinds of clothes we wear and what types of houses we live in. We depend on heat for life, for if the sun should cool, all life on Earth would disappear.

With this in mind, we should learn all that we can about heat. The study of heat in physics is called thermodynamics.

Step on a Crack

Have you ever wondered why sidewalks are laid out in sections with spaces, or cracks, between them?

You need:

an empty can
a large nail
a hammer
a lamp (100-watt bulb) with foil shade
a kitchen timer
pliers or tongs

What to do:

Hammer the nail into the bottom of the can. Work the nail in and out a couple of times to make sure that it slides through easily. Pull the nail out of the hole.

Turn on the lamp with the foil shade. Set the kitchen timer for two minutes. Now, using pliers or tongs, hold the nail over the lamp shade on the bulb

and heat it until the timer bell rings. Be careful not to touch the hot bulb or foil shade with your hands.

Try to put the nail back into the hole in the can.

What happens:

The heated nail does not fit back into the hole.

Why:

Heat from the light bulb excites the tiny, separate particles that make up the nail. These excited molecules move faster and spread out to take up more space. So the heated nail is now bigger than it was before and no longer fits in the hole.

Like the nail, a sidewalk's molecules also spread out on hot summer days. If there were no cracks between the sections of sidewalk, the heated molecules would have no room to expand and the concrete sidewalk would crack or break up.

HOT!

Day and Night in a Can

Why do you suppose the people who design clothing use dark colors for winter coats but white or light colors for summer things?

You need:

a lamp (100-watt bulb) with foil shade
a small can
black paint
petroleum jelly
2 pennies
a cotton swab

What to do:

Prepare for this experiment by painting one side of the *inside* of the can black. (Be careful of sharp can edges.) Leave the other half shiny.

When the paint is completely dry, take the cotton swab and put two dime-size blobs of petroleum jelly on the outside of the can, one in the center of the dark half and one on the other side. Push a penny down firmly into each blob so that it sticks. Make sure that one penny is stuck outside the dark half, and one outside the shiny half.

With the foil shade in position on the bulb, carefully balance the can, open side down, on top of the shade. Now, turn on the lamp.

What happens:

The petroleum jelly melts and the pennies drop off, of course, but the penny outside the darkened half *drops first*.

Why:

Although heat generated by the light bulb moves towards and into the can equally, the dark-colored surface soaks up more heat than the light or shiny surface. The petroleum jelly outside the black-painted side of the can melts faster, so the penny on that side usually wins the race.

Dark colors not only retain heat better, but they also do a better job of soaking up, or absorbing, the waves of heat that come to us in the light rays from the Sun. So, in winter, you want a dark coat to keep you warmer. Light-colored clothing reflects, or bounces, most of the heat-making light waves back into the atmosphere, so it keeps you cooler on hot, summer days.

HOT!

Decorate a Hat

How does heat change solid forms to liquids? Here's a chance not only to understand the process of melting, but to use it to make a fancy hat as well.

You need:

a lamp (100-watt bulb) with foil shade
a box of birthday candles (different colors)

What to do:

Before turning on the lamp, turn up the edge of the foil shade all the way around, giving it a nice big brim. Fit this hat-shape securely over the light bulb.

Turn on the lamp. Carefully, hold the tip of the birthday candle against the top of the foil hat.

What happens:

After a few seconds, the candle begins to change into a liquid that quickly drips away from the wick.

Why:

The heat from the light bulb excites the tiny particles, or molecules, of wax that make up the candle. As these molecules expand, or spread out, they begin to move about so wildly that they change into a liquid.

This process is, of course, called melting, and the temperature at which different substances melt is called their melting point.

What now:

Finish decorating the foil hat by letting the different-colored candles drip all over the hat and around the brim. When you are finished, turn off the lamp and let the hat cool before removing it from the bulb.

Later, if you are really in a fun mood, design a face on an apple or a potato and create "a someone" to wear the decorated hat.

Melt a Kiss

Our Earth gets its warmth from the sun's light. Fortunately, the Earth is constantly turning on its axis and is not made of chocolate.

You need:

a small paper bag
a chocolate kiss
some water
a lamp (100-watt bulb) with foil shade

What to do:

Put a little water into the bag. Swirl it around to coat the inside of the bag and pour the rest out. Put the chocolate kiss into the bag. Then, carefully, hold the bag over the shaded light bulb. Do *not* let the bag touch the shade.

What happens:

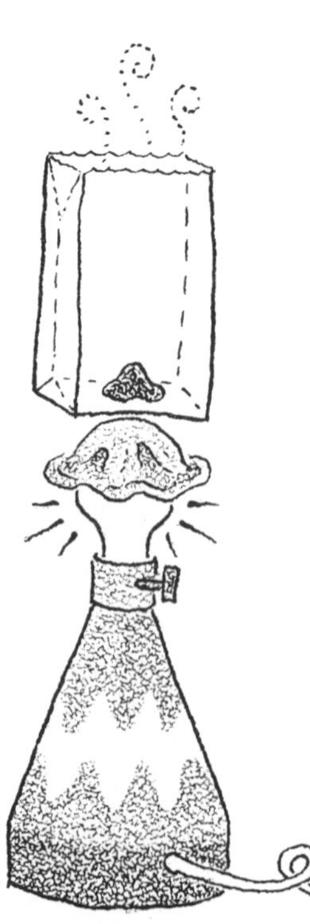

The chocolate kiss inside the bag begins to melt, but the bag doesn't get hot.

Why:

The heat energy coming from the light waves is soaked up, or absorbed, by the film of water molecules inside the bag and by the chocolate kiss.

This heat energy, which cannot be seen but can be felt, moves in circular waves away from its source, like the expanding ripples a stone makes when it is thrown into calm water. It is this form of heat, called radiation, which allows the sun to heat our earth, and the lamp to melt the chocolate kiss.

Color Me Warm

In houses and buildings with central heating systems, warmed air rises, cools off, and sinks down. This experiment uses water to give you a clear picture of how central heating works.

You need:

2 clear glasses
hot tap water
very cold water
food coloring
small measuring spoon

What to do:

Fill one glass with hot water, and the other with icy-cold water. Measure some food coloring into each glass.

What happens:

The food coloring quickly swirls around to mix with the hot water—but not so in the cold water.

Why:

Because its tiny, separate particles, or molecules, are moving faster than the cold-water molecules, the hotter water at the bottom of the glass whirls around and up to the top taking the food coloring with it. As the water at the top cools, warmer water takes its place, and the now cooler water sinks. This process, called convection, will continue until all the water is at the same temperature.

Because there is less motion in the cold water, the food coloring sinks to the bottom of the glass and stays in place longer.

Birds on a Wire

If you have ever grabbed a hot, metal handle of a cooking pan, you have actually felt hundreds of speeding molecules striking your hand like tiny bullets. Unfortunately, such a "hands on" experience usually results in a painful burn or blister.

You need:

a paper clip or stiff wire
a spring-clip clothespin
petroleum jelly
a small spoon
a lamp (100-watt bulb) with foil shade
a piece of waxed paper

What to do:

Straighten out the paper clip to make a straight piece of wire. Clip the clothespin to one end of the wire as a handle. With the spoon, line up three, separate, jelly-bean-size blobs of petroleum jelly on the

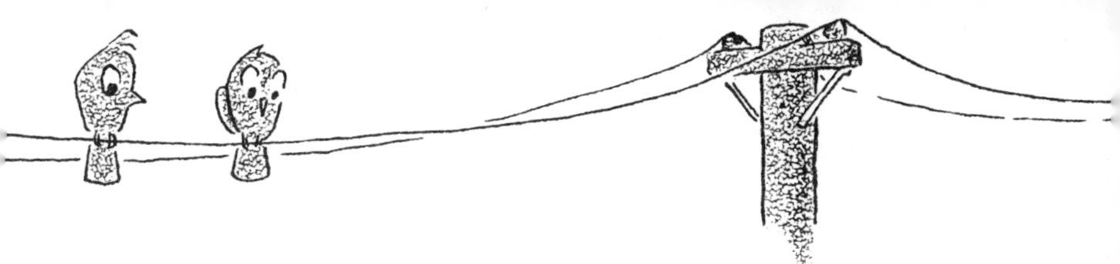

wire. Try to space the blobs evenly, so that they look like birds sitting on a telephone wire.

Place the end of the waxed paper underneath the lamp's base. With the foil shade covering the bulb, turn on the lamp. Holding the clothespin, touch the far end of the wire to the bulb and hold it there. The "birds" need to be sitting on the wire over the waxed paper leading away from the bulb.

What happens:

Beginning with the "bird" perched nearest the light bulb, one after the other begins to drop off the wire, until all three are gone.

Why:

When one end of the wire is heated by the light bulb, the tiny, separate particles, or molecules, in that area of the wire zoom into motion.

Soon these heated molecules begin to jostle the ones next to them, then those molecules bump others nearby, until the heat is passed all the way down the wire, melting each "bird" in turn.

This passing of heat from one excited, jostling molecule to another, like a handshake sent down a line or around a circle, is called conduction.

WHY IS THERE AIR?

Air alone has no color, taste, or smell. You can't see it, and you can put your hand right through it without feeling anything. However, air is more than just nothing at all. It is actually made up of many gases, mainly nitrogen and oxygen, which consist of tiny molecules that are far apart and move about quickly. This is why gases are thin and appear invisible.

Scientists estimate that 1 cubic inch (16.4 cubic centimetres) of air contains about 300 billion billion molecules! Even though the molecules are so tiny, there is still plenty of space between each one. Each air molecule has enough energy to zip through space at 11,000 miles (1600 kilometres) per hour.

The Collapsing Bottle

The following experiment will help your recycling efforts by giving you more room in the collecting bin.

You need:

a plastic 2-litre bottle with cap
very hot tap water

What to do:

Pour hot water into the bottle until it is about half full and swish it around for about a minute. Then pour the water out and, quickly, put the cap on and twist it tightly.

What happens:

The sides of the bottle suddenly collapse inward!

Why:

The hot water heats the air inside the bottle and, with the cap left off, it fills to the brim with warm air.

When the hot water is poured out and the cap is replaced, the air inside of the bottle quickly starts to cool. Since cooler air takes up *less* space than the same amount of warmer air, there's now extra room in the bottle!

To fill that extra space, the sides of the bottle are pushed in by the force of the air pressure outside the bottle, which is constantly pressing in every direction.

The Wonderful Whistle-Stick

Can you turn a piece of wood into a whistle? Sure you can. It's a great experiment, and fun, too.

You need:

a wooden paint stirrer or small paddle
a hammer   a large-size nail
a long piece of string

What to do:

Use the hammer and nail to make a hole in the narrow end of the stirrer or paddle. Put one end of the string through the hole and tie a tight knot.

Now, make two or three holes in the wider end of the wood. You can put the holes all in a row, or make up your own pattern.

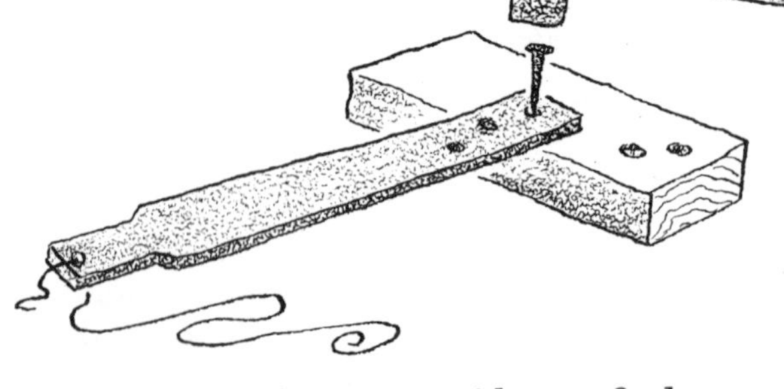

To hear your whistle-stick, go outside or find a large open area where you can swing the stick without breaking anything. Hold tightly onto the loose end of the string and whirl the paddle around in front of you or over your head.

What happens:

You hear an unusual whistling sound over and over again.

Why:

As you whirl the paddle around, the air passes through the holes in it at a higher speed than the air going around the paddle. When this happens, the paddle whistles.

What now:

Different numbers and sizes of holes make different whistle sounds. You might want to make several whistle-sticks—some with only a few small holes to catch the air and others with larger or a lot of holes. Then you can compare the whistle sounds that each one makes.

The Talking Coin

You may have heard somebody say that money talks, but until you do this experiment you have probably never actually seen it speak.

You need:

a plastic 2-litre bottle
a quarter
a cup of water
a freezer
a kitchen timer or watch

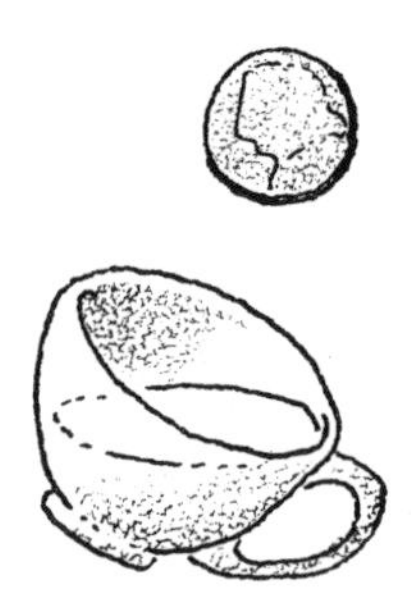

What to do:

Put the quarter in the cup of water and place the empty bottle in the freezer for five minutes.

When the time is up, remove the bottle from the freezer and, immediately, cover the mouth of the bottle with the wet coin. (It is important to *completely* cover the bottle's mouth with the coin.)

What happens:

The quarter becomes a tongue for the bottle and begins to chatter at you.

Why:
When the bottle was put into the freezer, the air molecules inside of it cooled and moved closer together. Since the air in the bottle now took up less space, it left room for extra air to flow in—so it did.

When the bottle was removed from the freezer, however, the air molecules inside of it began to warm up and spread out again. It's a great example of, "There was enough room for everyone to sit comfortably in the car until we all put on coats and it was crowded." Suddenly there was no room for the extra air molecules.

It is that "extra air" that is being pushed out of the bottle as the air warms that makes the coin move up and down as if it were talking.

Launch Your Own Astronauts

In the same way that a hurricane can blow you off your feet, you can make a flying chamber for your own band of brave adventurers.

You need:

Ping-Pong balls
a blow dryer
permanent markers (optional)

What to do:

For fun, using the markers, make a face on each of the balls. You could even identify your astronauts by writing their names on the balls.

Next, plug in the hair dryer, turn it to high-blow, heat setting, and point it *straight up*.

Place one of your homemade astronauts in the dryer's airstream and let go.

What happens:

The astronaut is launched towards the ceiling, but stops and bounces around in the airflow partway up.

Why:

The airstream from the dryer pushes the astronaut upwards, against the force of gravity, until the upward and downward pushes are equal and the astronaut just floats.

The high pressure in the still air surrounding the airstream keeps the astronaut in the center of the invisible flying chamber.

What next:
Try letting two or more astronauts fly at the same time. You may be able to do this if the blow dryer's airstream is wide enough (or you have an attachment called a diffuser that spreads out the airflow). If not, your astronauts will probably "bounce" off each other—into unknown galaxies.

The Incredible Shrinking Face

Air has a magical quality about it, in that it can expand or shrink flexible material almost instantly.

You need:
a light-colored latex balloon
a permanent marker

What to do:
Blow up the balloon, until it is fully inflated, then hold it tightly in one hand so that no air leaks out.

While holding the balloon, pick up a black or dark marker with your other hand. Draw a large face on the balloon, completely covering one side of it.

Next, relax your grip on the balloon's neck and watch as you slowly let a constant stream of air escape.

What happens:
Right before your eyes, the huge face that covered the whole side of the big balloon shrinks to a miniature drawing.

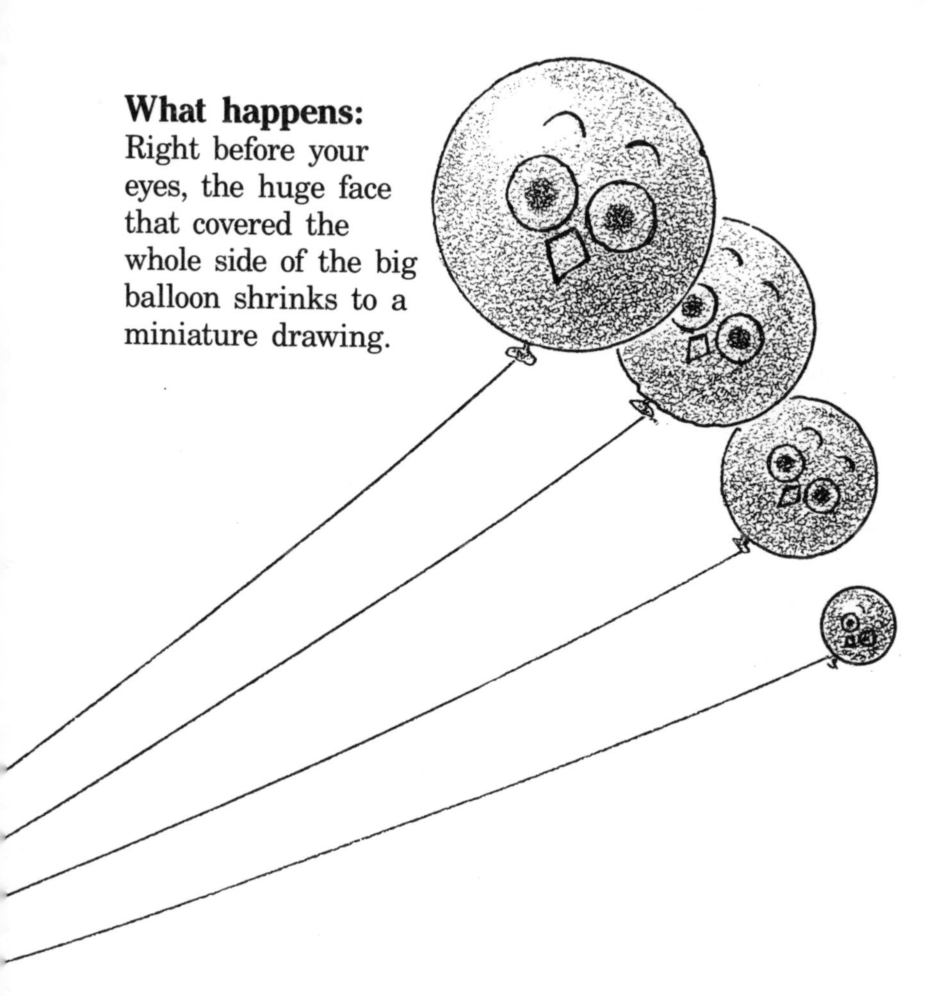

Why:
As the air is allowed to escape, the balloon material that expanded as you blew into the balloon goes back to its original small size, taking the marker "face" with it.

If your marker ink tends to smear, spray the drawing while the balloon is full of air with a clear gloss aerosol finish to keep it neat.

The Trick Straw Race

This race is just for fun, but you might want to challenge someone you *know* is a good loser to compete.

You need:

2 identical glasses, evenly filled
2 striped straws a straight pin

What to do:

Before the race, use the straight pin to punch fifteen or twenty small holes toward one end of the "trick" straw. Punch the holes where the colors change on the straw so that they will be harder to see.

Place the straw, punctured end up, in your friend's drink and say, "Let's see who can drink it all first."

What happens:

While your glass empties quickly, most of your friend's drink will remain in the glass.

Why:

By sucking on the straw, you are lowering the air pressure inside of it, so the air pressure pressing down on your drink pushes it up the straw and into your mouth. In the "trick" straw, air rushes into the holes in the straw so that your friend can't lower the air pressure inside—at least not enough to win the race.

The Collapsing Tent

Everybody knows that moving air has more power than air that is not moving. Or does it? The following experiment will help you draw your own conclusions.

You need:

a small sheet of paper
a table or countertop

What to do:

Fold the paper in half, creasing it with your finger to form a tent.

Open the tent and place it near the edge of the table or a countertop, with one open side facing you.

Get into position so that your mouth is level with the edge of the table, take a big breath and blow a *steady* stream of air *through* the tent.

What happens:

The tent collapses, becoming a flat sheet again.

Why:

When you blow through the tent, you are lowering the air pressure inside of it. This allows the higher air pressure above the tent to crush down and flatten it.

Make a Parachute

Race car builders make their creations as streamlined as possible so that these machines cut through the air and attain the highest speeds. Parachute makers, however, use yards and yards of billowing fabric so that their creations will grab as much air as possible on the way down.

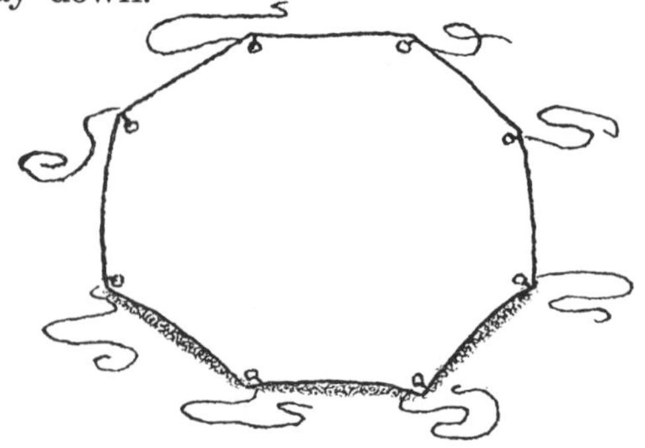

You need:

a lightweight plastic grocery bag
scissors
a nail
8 pieces of yarn (same length)

What to do:

Cut a large square piece from the *front* of the bag (without a seam). Trim off the corners to form an octagon, or 8-sided shape. With the nail, carefully punch a small hole near each angle of the plastic so that you have 8 holes, spaced evenly, all the way around.

Tie one end of the pieces of yarn securely to each hole. Now, pull the other ends of the 8 pieces of yarn together and tie them in a tight knot.

Push the nail through a few of the knotted strands as a weight—this is your parachutist.

To test your parachute, stand on a sturdy chair or go outside in search of a breeze. Using both hands, hold the parachute out in front of you, high above your head, and release it.

What happens:

The plastic parachute billows out and floats gently downward and away from you.

Why:

It is a law of physics that the larger the surface area in contact with the air, the harder it is for that object to travel through the air. So, the larger the parachute, the slower it moves—which is what you want if you are falling from a height!

But, the umbrella-like shape over a weight, like a truck parachuted out of a plane, also traps air beneath it. It is the air spilling off to one side of the parachute's top, or canopy, that makes it drift away from you.

What now:

Make a *very small* hole in the middle of the canopy. This will make the parachute fall straighter. Now you can mark out a landing area and try to have your parachutist hit the target.

The Singing Balloon

Most balloons are content to just float around quietly. This mezzo-soprano is dying to let loose and sing.

You need:

a balloon
some music

What to do:

Blow up the balloon, making it as full as possible.

Pinch the neck of the balloon. Pull your hands apart gently to release a slow, steady stream of air from the balloon. Move your hands back and forth, stretching the balloon's neck to the music's beat.

What happens:

A high-pitched squealing sound is heard. It changes tone as you pull and release the neck of the balloon.

Why:

The molecules of air packed inside the balloon rub against the rubber molecules of the balloon's neck as they rush by on the way out. This causes the rubber in the balloon's neck to shake, or vibrate, and makes the squealing noise. Stretching the neck of the balloon makes it vibrate at different speeds to make different sounds.

For fun, why not form a Singing Balloon Band.

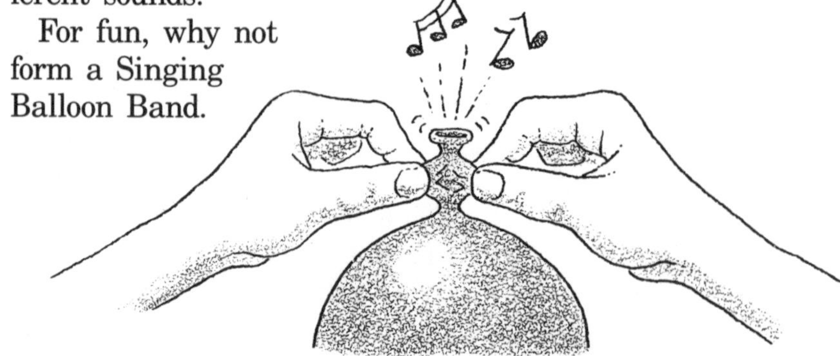

The Rising Notebook Trick

Even though you might think of air as a gentle, invisible "free spirit," it can be quite strong and powerful when pressured.

You need:

a school notebook
a medium-size balloon
a table or countertop

What to do:

Place the balloon at the edge of the table or countertop with the mouth of the balloon sticking out towards you. Put the notebook on top of the balloon. Hold the balloon by the neck and blow into it.

What happens:

The notebook rises.

Why:

The high air pressure from your lungs causes the balloon to expand and lifts the notebook off the table. In repair garages, whole cars are lifted up in a similar way by air pressure.

Unfortunately, the air pressure in the balloon does not have a lifting effect on the grades of any homework in the notebook.

Air-Head Person

Even though you cannot actually see air, you can trap it inside something and see the shape that it makes. You can also bring that shape to life—sort of.

You need:

a latex balloon
a square of heavy cardboard
a thick nail
cellophane tape
scissors
permanent markers

What to do:

Blow up the balloon and tie a tight knot in the neck.

What happens:

Your once-limp balloon has taken on a round, roly-poly shape.

Why:

The air you blew into the balloon with the pressure of your lungs is now trapped in there. Crammed by the balloon into the smallest possible shape, it pushes equally against all sides of the balloon, making it round.

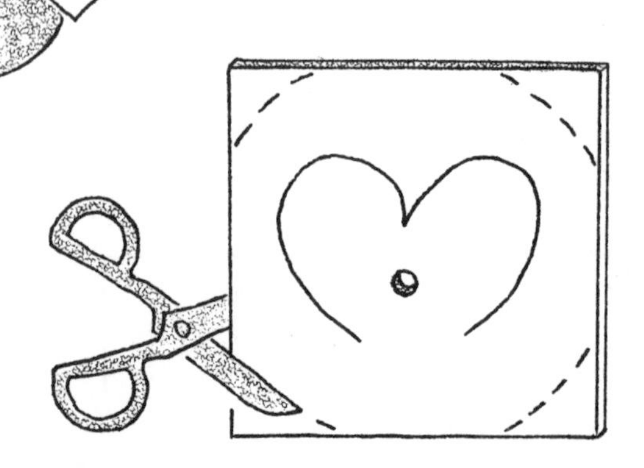

What now:

Turn that roly-poly shape into an air-head person. Draw a large rounded ω on one side of the cardboard to make the feet. Cut around the edges of the cardboard, and then punch a hole in the center of it with the nail.

Carefully poke the knotted end of the balloon through the hole in the cardboard feet. Pull the knot forward, towards the "toes," and tape the knot securely to the cardboard with some tape.

Your air-head person is now ready for a face (eyes, ears, mouth, nose, hair, glasses, mustache, or whatever you like), using the markers. And don't forget to print a name on its cardboard feet.

Decorate your room or make lots of air-heads for a birthday party, depending on how much air you have to spare.

Real String Soap in a Bottle

By knowing a little about how air pressure works, you can make a great trick that is sure to surprise everyone.

You need:

a plastic detergent bottle with cap
white string
scissors

What to do:

Remove the cap from the detergent bottle. Using the scissors, bend the plastic strip inside the cap to one side, or remove it altogether. (If you have trouble doing this, ask someone to help you.)

From the inside of the cap, poke one end of the string through the holes and out the top. Make sure that the string glides easily back and forth through the openings.

Then, tie a fat knot in each end of the string so that it cannot pull through the holes and will stay in place. Leaving only the fat knot outside, covering the cap opening, hide the rest of the string down inside the bottle. Now, replace the cap.

To do the trick, point the bottle and give it a quick squeeze.

What happens:

Watch it! A stream of white "string soap" squirts out!

Why:

When the bottle is squeezed, air rushes out through the holes in the cap and the top shooting the knot covering them into the air. Also, the air pressure inside the plastic bottle, when you squeeze it, is greater than the air pressure outside of it. This difference in pressure also helps to force the string out.

Underwater Eggspert

You probably know that eggs have yolks, but did you know that they also contain air? This *eggs*periment will show you how to prove it.

You need:

a fresh egg
a deep cereal bowl
hot tap water
yellow food coloring (optional)

What to do:

Carefully, place the egg in the bottom of the bowl and fill the bowl with hot tap water. Quickly add a little yellow food coloring.

Watch the egg closely for several minutes.

What happens:

Streams of tiny bubbles rise to the top of the water from the submerged egg.

Why:

When the air inside the egg is heated by the hot water, the air molecules expand. Many of the now crowded air molecules push their way out of the egg through some of the almost 7,000 openings, or pores, in the egg's shell. These heated air molecules exit the egg, usually without cracking the shell, and rise to the water's surface as bubbles.

Acupuncture Balloon

Acupuncture is an ancient Chinese medical procedure often used to relieve pain. You can practise acupuncture, without a medical license, on a balloon.

You need:

a small latex balloon   a string
adhesive or other strong sticky tape
5 or 6 sharp straight pins

What to do:

Blow up the balloon about three-quarters of the way. Knot the end and tie the piece of string around the neck to help hold the balloon.

Cut off 5 or 6 pieces of the strong sticky tape. Press the pieces as evenly as you can around the outside of the balloon. Make sure that each piece of tape is secured *tightly* to the balloon.

Now, one by one, carefully stick a straight pin through the middle of each piece of tape and into the balloon, puncturing the balloon all over.

What happens:

Nothing! The balloon doesn't burst!

Why:

When the pins pass through the tape, the sticky adhesive on it forms a seal around each pin that prevents the air from escaping from the balloon when you push them in. And, as you may know, a balloon only "pops" when the air is under pressure, and it is suddenly allowed to escape.

"Boil, Boil, Magical Water"

Would you believe you can boil water without using a stove? Here's the key to this old, well-kept secret.

You need:

a clear drinking glass
  (a narrow one is easier
  to handle)
water
a handkerchief-size
  square of cloth
a rubber band
a sink

What to do:

Fill the glass about half full of water. Lay the cloth evenly over the top of the glass and push the center of it down into the water.

Then, put a rubber band tightly around the top to hold the cloth edges against the sides of the glass.

Turn the glass upside down over the sink. Some of the water may dribble out, but most of it will stay inside the glass.

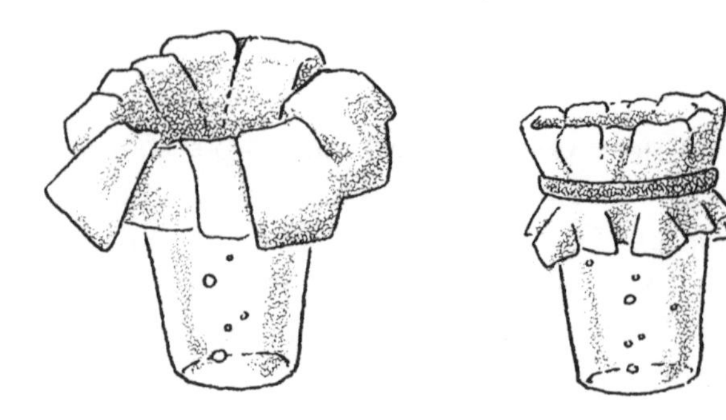

Hold the cloth tightly around the neck of the glass, between the rubber band and the covered opening, and push down hard on the upside-down bottom of the glass.

What happens:

The water starts to boil! (It may take a couple of tries to get the hang of this, but don't give up.)

Why:

Of course, the water isn't really boiling, because there is no heat source. Actually, it is the air that comes in through the cloth when the water is squeezed out (by pressing on the bottom of the glass and tightly pulling on the cloth) that causes the bubbles—and makes it look as if the water in the glass is boiling.

What now:

Once you can control the bubbling, use this experiment as a trick lie detector. Ask friends some questions and tell them that the water will boil if they lie, but won't if they tell the truth.

NOTE: You can make this trick look more mysterious by tinting the water in the glass with food coloring.

WATER, WATER, EVERYWHERE!

Water is the most common substance on Earth, covering more than 70 percent of its surface. Although it sometimes appears blue or green, water is a clear, colorless liquid. The thing about water that makes it different from most other liquids is that it is lighter as a solid (ice) than as a liquid. Water is necessary to sustain life, making up most of an animal's blood, a plant's sap, and two-thirds of the human body.

Water is also a large part of our environment. It occurs as rain, sleet, snow, hail, frost, fog, dew, steam, humidity, and clouds. Not only are there lakes, rivers, oceans, and swamps that cover three-fourths of the earth's surface, but water also accumulates in spaces between and within rock beneath the earth's surface, supplying wells and springs and sustaining streams during periods of drought.

Water circulates constantly through our world, being used but never being used up. The glass of water that you drink today could contain the very same water molecules that a thirsty caveman enjoyed many thousands of years ago!

"I Was Here First!"

This experiment proves that two forms of matter cannot occupy the same place at the same time.

You need:

a clear drinking glass
several marbles (any color)
water
masking tape

What to do:

Fill the glass to about half full of tap water. Next, put a piece of masking tape on the outside of the glass to mark the water level.

Now, tilt the glass and carefully slide the marbles, one by one, down inside the glass to the bottom.

Set the glass upright and check the water level.

What happens:

The water level is higher than it was before.

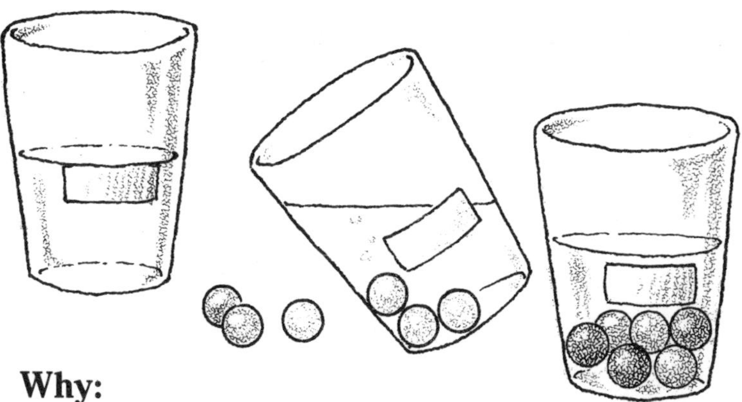

Why:

The water and the marbles are both examples of matter that cannot share space. When the marbles are added to the glass, they are heavier than the water so they roll to the bottom of the glass and push the water there out of the way. The water level is, therefore, pushed up above the masking-tape marker.

Flowing Fountain

Blaise Pascal, a French physicist in the 1600's, discovered the scientific principle of how pressure affects liquids. Every time you enjoy watching water "dancing" from decorative fountains, you should thank Pascal.

You need:

an empty can
a hammer
a nail (birthday-candle size)
freezer tape
water

What to do:

Using the hammer and nail, punch 8 holes evenly around the can, about two finger digits up from the bottom rim.

Next, make a second row of holes, about one finger digit above the first row, except this time make only 4 holes. Try to space them evenly around the can in a nice pattern.

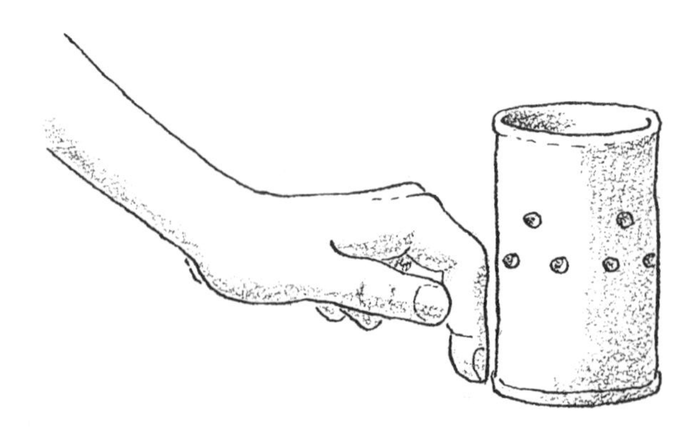

Tear off two strips of freezer tape, each long enough to reach all the way around the can, and carefully cover each row of holes with the tape.

Fill the can with water, hold it over a sink basin, and rip off both tapes together. (You may need an extra helping hand here.)

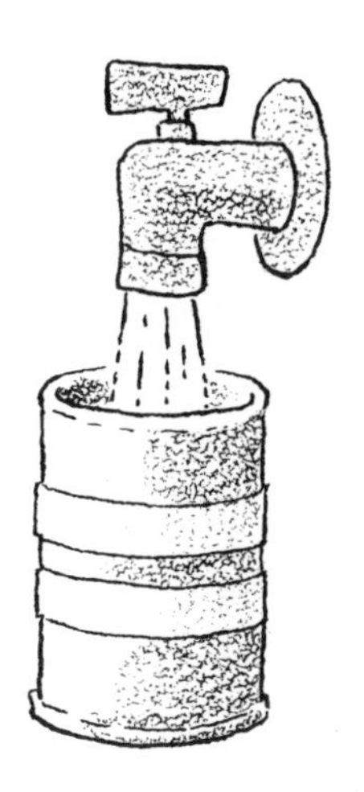

What happens:

The water flowing from the bottom holes squirts out just a little bit faster and farther than the water from the upper row of holes.

Why:

The water in the lower part of the can is under more pressure, from the weight of the water above it, than the water higher up in the can.

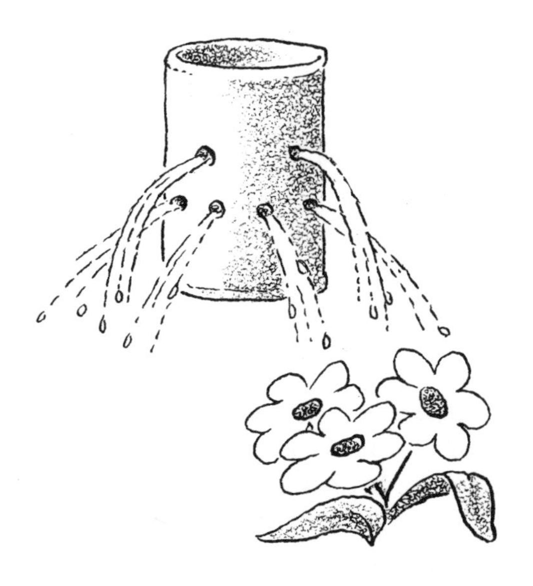

Two Water Towers

Water towers come in all sizes, but does size make a difference to the water? Let's see.

You need:

2 cans (one tall and thin;
one short and bigger around)
a hammer
a nail (birthday-candle size)
freezer tape
water

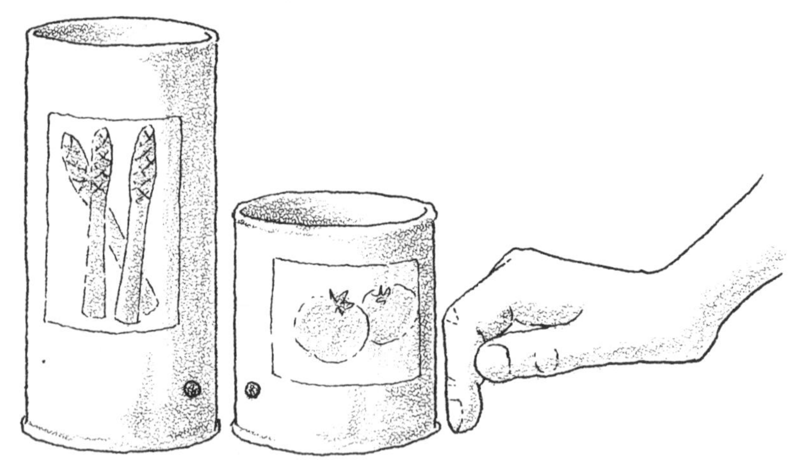

What to do:

Using the hammer and nail, make a hole about one finger digit up from the bottom rim of each can. Be sure to make the holes identical. Cover each hole completely with a small strip of freezer tape.

Fill each can to the brim with water. Set both cans on the edge of the sink, with the holes towards the basin. Rip off the tapes.

What happens:

The stream of water that flows out of the taller, thinner can is longer than the stream of water from the shorter can.

Why:

It is the depth of the water that determines how fast the water flows out of the hole, and the deeper water is in the taller can. Shorter towers that are bigger around might hold as much or more water, but that water will come out slower and with less power. That is because the weight of the water pressing down, or water pressure, is less.

"I Think I'll Eat Worms"

When this experiment with bringing spaghetti to life is finished, you can amaze your friends even more by eating the "worms." They might taste a bit like pickle.

You need:

a few strands of cooked spaghetti
a deep glass bowl or large jar
a cup of clear vinegar
a cup of tap water
red and blue food coloring (optional)
2 tablespoons of baking soda (30mL)

What to do:

Tear the strands of cooked spaghetti into several worm-size pieces.

In a bowl or jar, mix the cup of vinegar and the cup of water.

Next, add 3 drops of red food coloring and 3 drops of blue food coloring to the mixture and stir to make the color purple.

Slowly add the 2 tablespoons of baking soda, then drop in the pieces of cooked spaghetti.

What happens:

The purple "worms" seem to come to life! They swim back and forth, rising to the top of the water and then falling back to the bottom of the container.

Why:

When vinegar and baking soda mix, they form tiny gas bubbles. These bubbles attach themselves to the small strands of spaghetti, raise these "worms" to the top, and then burst. The pieces then fall back to the bottom where, if more gas bubbles attach themselves, the purple "worms" will continue to swim up and down in the bowl.

The Power of Water

Turn on a faucet in your home until a stream of water trickles out. Then, put your hand directly under the faucet, and try to stop the water from coming out. You can't?

This experiment will help you understand why you have to turn off the faucet to stop even the littlest stream of water.

You need:

a large-size can
a hammer
a nail (birthday-candle size)
freezer tape
water
sink

What to do:

With the hammer and a nail, punch three holes into one side of the can. Put one hole near the bottom of the can, just above the rim. Make the other two holes about a finger-digit apart, going up the side of the can, above the first hole.

Cover all three of the holes with one long strip of freezer tape, then fill the can with tap water.

Place the can on the edge of a sink, with the holes positioned over the basin, and rip off the tape.

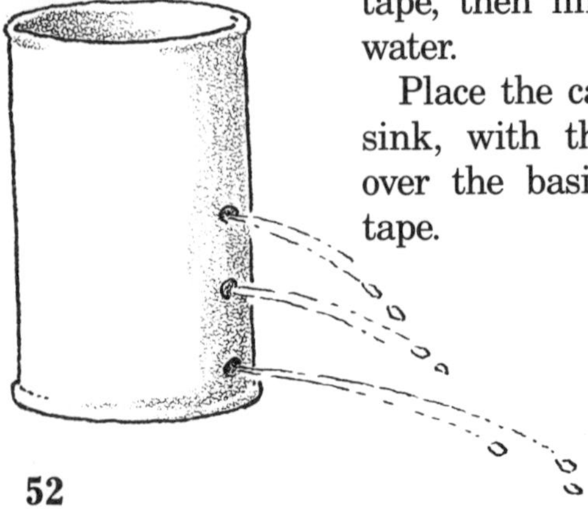

What happens:

The water stream from the lowest hole at the bottom of the can is the longest. The water stream from the middle hole is next longest, and the stream from the highest hole is the shortest.

Why:

The water at the bottom of the can is under more water pressure than the water above it. The more pressure there is, the longer the stream of water.

Your faucet is like the long stream from the lowest hole. Many cities pump water into tall tanks, or towers. The pressure of the water from these tall containers forces water through long pipes beneath the ground and into people's homes. The water from the faucet has all that pressure behind it, that's why you can't stop it with just your finger or hand.

You didn't really think that all that water was just right inside your wall, did you?

Ice Boat Float

What strange thing about water protects hundreds of millions of fish in lakes and rivers each winter? Here's how to find out.

You need:

a large jar filled with very cold water
an ice cube

What to do:

Place the ice cube in the jar of water.

What happens:

The ice cube floats like a boat.

Why:

When water molecules freeze, they expand and spread out. This means that ice is not as heavy or dense as water. Because frozen water is lighter than regular water, it floats!

This lucky law of nature causes water to freeze from the *top down*. When the layer of water on the surface turns to ice, it prevents the water below it, where the fish live and swim, from freezing, too. So, in winter, the fish's watery world is protected by a floating "sky" of ice.

The Floating Glass

Floating clouds, yes; but floating glasses!

You need:

2 glasses that fit inside each other (glass works best)
water

What to do:

In one drinking glass, pour in just enough water to cover the bottom. Then place the second glass inside the first.

What happens:

The inside glass floats. (If this does not happen, add a little more water and try again.)

Why:

The amount of water in the bottom of the first, outer, glass is heavier than the weight of the inside glass.

It was Archimedes, a Greek mathematician who lived over 2,000 years ago (287–212 B.C.), who discovered that a floating object displaces, or pushes out of its way, an amount of liquid equal to its own weight.

Disappearing Salt

If you have ever gone swimming in the ocean, you have probably tasted salt in the water. No matter how hard you looked, though, you couldn't see it. Why not?

You need:

tap water
a clear drinking glass
a straw
a box of salt
a measuring cup

What do to:

Fill the glass *right to the brim* with warm tap water. Measure a half cup of salt and, very slowly, pour it into the full glass of water, while stirring gently with the straw.

What happens:

If you pour carefully, you can add the entire half cup of salt to the full glass of water without any of the water overflowing.

Why:

The water does not spill over when the salt is added because no extra room is needed. The molecules of water have spaces between them. These spaces are filled nicely by the molecules of salt, just like cars driving into garages that fit them. Such a neat arrangement between two substances is called a solution.

“Freeze Me and I’ll Burst!”

Have you ever had the water pipes in your home burst because water froze inside them? The following experiment explains why smart people allow faucets to drip, drip, drip, on freezing winter nights.

You need:

a small jar   tap water   a square of cardboard

What to do:

Fill the jar to the brim with tap water. Cover the jar’s top completely with the cardboard square.

Now, carefully place the jar in the freezer and wait until the water freezes.

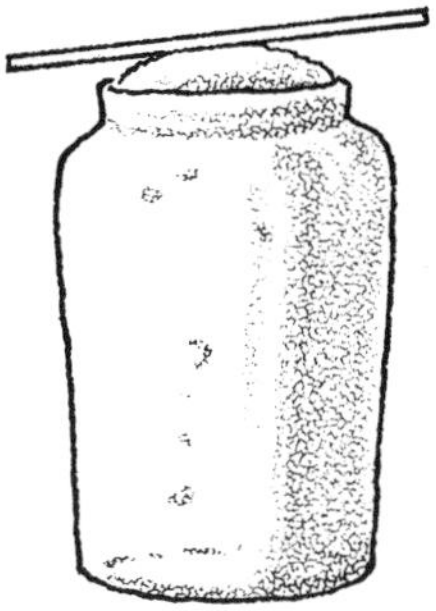

What happens:

The frozen water lifts the cardboard square above the top of the jar.

Why:

When the temperature of water drops below 32 degrees Fahrenheit, or freezes, its molecules spread out and need more space—just like when water is heated (water is strange in this way). The freezing water molecules push up out of the jar looking for room to spread out.

If, instead of just laying a loose piece of cardboard on top, you had put a tight-fitting lid on the jar, what would have happened? The freezing water molecules would have expanded and broken the jar. This is what can happen to your water pipes in winter, if some of the water molecules are not allowed to escape from the faucet.

1 + 1 Does Not Always = 2

You might be a good math student, but you will have to be a good physics student to figure out this experiment.

You need:

a large-size glass jar
masking tape
a pen
a cup of sugar
a measuring cup
a paper towel
a drinking straw
warm water

What to do:

Place a strip of masking tape down the outside of the jar.

Pour one cup of warm water into the jar and mark the level that it reaches on the tape. Then, add a second cup of warm water and, again, mark the water level on the tape.

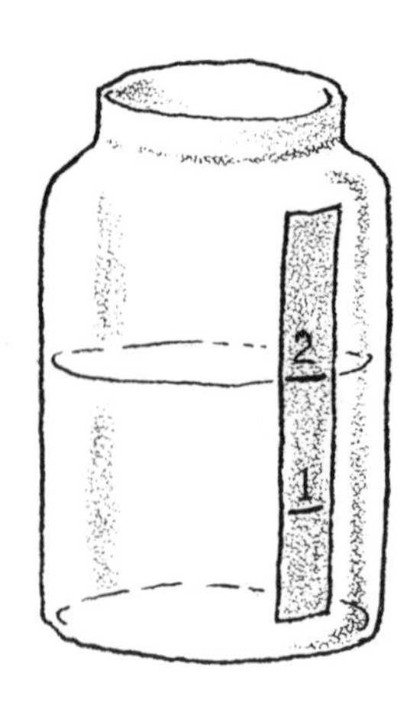

Empty all of the water out of the jar and dry the inside of it with a paper towel.

Now, pour one cup of warm water into the jar. Follow that with one cup of sugar. Stir this solution well with the straw and then check the liquid level on the masking-tape measuring strip.

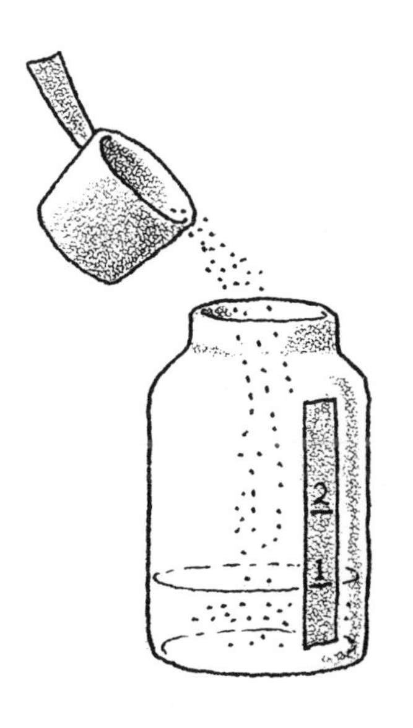

What happens:
The liquid level of one cup of water plus one cup of sugar does not reach the two-cup mark of the tape.

Why:
If you caught the clue word, *solution*, when you were instructed to stir the sugar and the water together, you probably already know the answer.

The substances in a solution fit neatly together, like puzzle parts. Instead of taking up their own space, the grains of sugar simply fill in the empty spaces around the water molecules to make something entirely new, a solution called sugar water . . . but less of it than you thought you would have when you added the sugar and water measurements together.

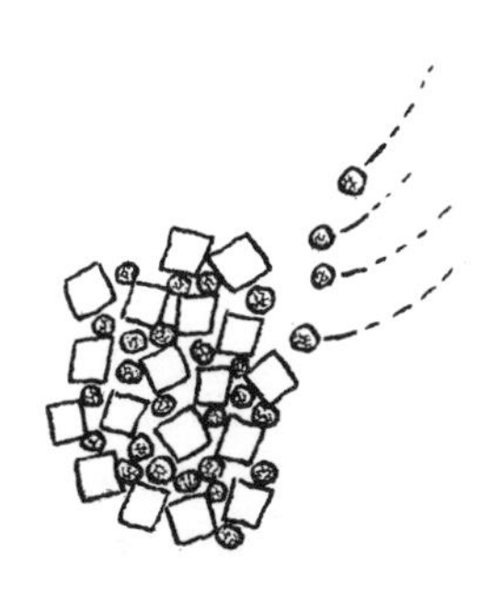

"Give Me Room!"

Just as you probably don't like to sit close to anyone when you are hot and sweaty, neither does a molecule of hot water.

You need:

a small glass jar
a pot or frying pan
water
access to a stove or hot plate (ask for permission or help)

What to do:

Put some water into the bottom of the pot and place it on the stove or hot plate.

Fill the jar *right to the brim* with water. Then, carefully so you don't spill the water, place the jar in the middle of the pot.

Turn the stove or hot plate on high heat and wait a few minutes as the water in the pot heats and begins to boil. Watch the water in the jar.

What happens:

The water in the jar overflows into the pot.

Why:

Like other liquids, water expands and needs more space when it is heated. As the water gets hot, the molecules in the jar start to bounce around rapidly, bumping against each other and looking for room to spread out, until they spill out over the top.

WARNING: Turn off the stove or hot plate and let the water cool before removing the jar from the pot.

The Shrinking Molecule

When you are very cold, you sometimes huddle up, trying to make your body smaller and smaller to keep warm. What does a water molecule do when it's cold?

You need:

a jar
tap water
access to a freezer
a kitchen timer or watch

What to do:

Fill the jar to the brim with tap water. Place the jar uncovered in the freezer. Set the timer for 30 minutes. When the time is up, take the jar from the freezer.

What happens:

The water level in the jar drops below the brim, even though none of the water has splashed out.

Why:

As the water's temperature gets colder, to about 39 degrees Fahrenheit, its molecules contract, or huddle closer together—maybe to keep warm. As a result of this "scrunching up," the molecules of cold water in the jar take up less space than they did before.

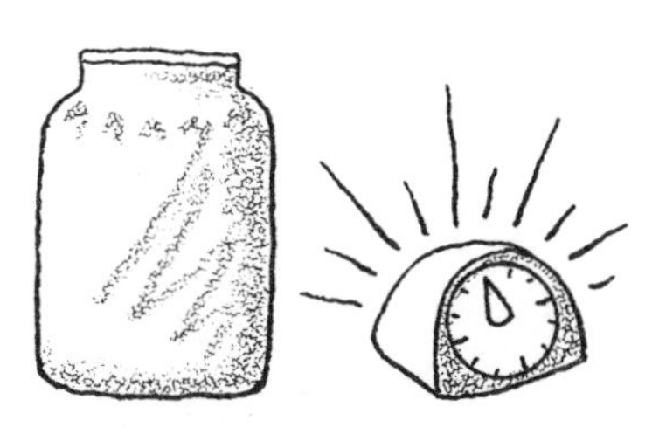

Shy Blue

Have you ever had a shy moment, such as at a family reunion, when you wanted to run and hide? If you have ever felt this way, then you can sympathize with Shy Blue.

You need:

a white plate
water
blue food coloring
rubbing alcohol
an eyedropper or straw

What to do:

Take some water and pour a small circle of it in the middle of the plate.

Put 3 or 4 drops of food coloring into the water.

Now, fill the eyedropper with alcohol and then squeeze it out, letting the drops fall against the circle of water. Do it again.

What happens:

Sky Blue trembles and shrinks back, away from the alcohol.

Why:

Both the water and the rubbing alcohol have something called surface tension, a thin, invisible "skin"

that holds them together. The surface tension of the water is stronger than the alcohol's, so it pulls its molecules *away* from the alcohol, trying to escape from its touch.

MAKE A "DROPPER"

If you don't have an eyedropper (or medicine dropper) for use in some of these experiments, you can use a straw. To do this, put the end of the straw into the liquid you need. Slowly and carefully, suck on the straw to pull a little of the liquid into it; then quickly put your finger over the top of the straw. Air pressure will hold the liquid inside. Next, move the straw into position and release your finger, and your "drops."

Square Bubbles from Square Holes?

Can it be done? Try the following experiment to see if you can blow a square soap bubble.

You need:

a piece of wire or pipe cleaner
dishwashing detergent
glycerin (optional, from drugstore)
tap water
a saucer

What to do:

Bend the wire or pipe cleaner into a square shape with a handle.

With your finger, mix some water with a few drops of dishwashing detergent together in the saucer. Adding 2 or 3 drops of glycerin makes bubbles last longer.

Next, dip the side of the square shape into the soapy solution. Be sure to completely cover the shape with the mixture.

Lift the square form to your mouth and blow gently.

What happens:

The bubble that emerges from the square form is *not* square, but round.

Why:

The culprit is surface tension—that attraction of molecules for each other that forms a "skin" over them.

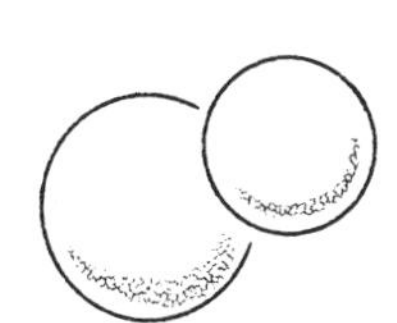

In this case, the surface tension of the soapy water making up the bubble pulls it into a rounded shape, no matter what shape you use to make the bubble. The attraction of the molecules makes a sphere, or ball, because that's the shape that allows them to be closest together.

More Than Enough

Is a cup of water full when it's full? Can it be more than full? Seeing is believing.

You need:

- a cup and saucer
- tap water
- an eyedropper or straw
- a table or countertop

What to do:

On a table or countertop, set the cup in the saucer and fill it to the brim with tap water. Next, draw some more tap water into the eyedropper. Add at least 20 drops of water, one after the other, to the full cup.

Then, bend over so that you are level with the cup and look at the water's surface from the side.

What happens:

Without spilling, the water rises over the rim of the cup like a large bubble.

Why:

When the water molecules across the surface are linked together with surface tension, they are strong enough to hold back the water so that it rises above the cup's rim without spilling.

At some point, if more water drops are added, the mound of water in the cup will become so high and heavy that the surface molecules will lose their grip on each other and will tumble out of the cup and into the saucer. Ouch!

Water "Glue"

While water and flour mixed together make a paste that will hold paper together, water "glue" (without the flour) is really a trick your eyes play on you.

You need:

a faucet with
running water
a large metal spoon

What to do:

Hold the rounded part of the spoon (bottom side up) under the stream of water.

What happens:

The water sticks to the spoon as if it were glued to it.

Why:

The reduced air pressure under the water on the spoon's rounded bottom holds the flowing water against the spoon, rather than allowing it to splash away.

But, it is the rapid movement of the running water, so fast that your eye can't follow the speeding molecules, that makes it look as if the water were "glued" to the spoon.

What now:

Turn the spoon over, right side up, and hold it under the stream of water from the faucet again. Does the water come unglued this time?

Stick Together, Stay Together

You often use water to wash off your sticky hands, but have you ever thought that water itself might be sticky? Let's find out.

You need:

a foam plastic cup
a pencil or nail
another cup or glass
water
a sink

What to do:

Using the pencil, punch two small holes in the bottom of the foam cup. Make the holes as close together as possible without allowing them to touch.

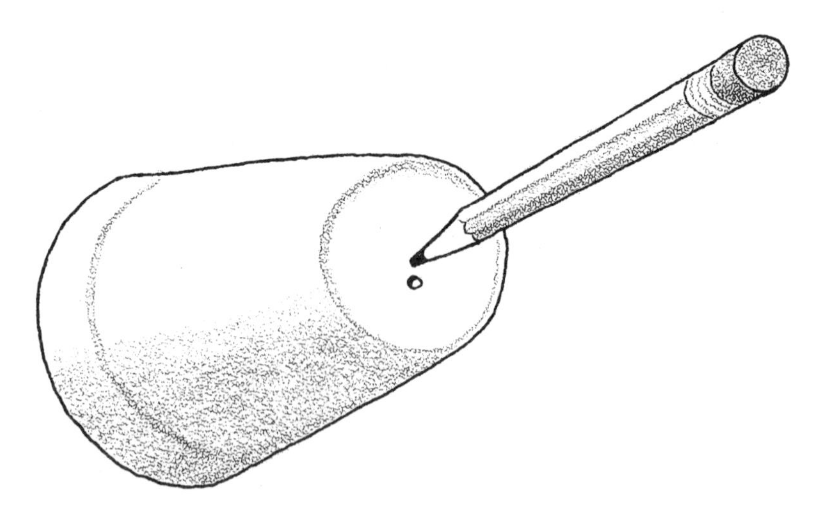

Next, fill the other cup with water, hold the cup with the holes up over the sink basin, and pour the cup of water into it. Now, very quickly, using your thumb and index finger, pinch together the two streams of water flowing out of the holes.

What happens:

The two separate streams of water come together and form one stream. (You might have to refill the cup a couple of times until you get the pinch right.)

Why:

Water molecules have such an attraction for each other that, when they come near, they grab on to one another and stick together. This sticking-together action is known as cohesion.

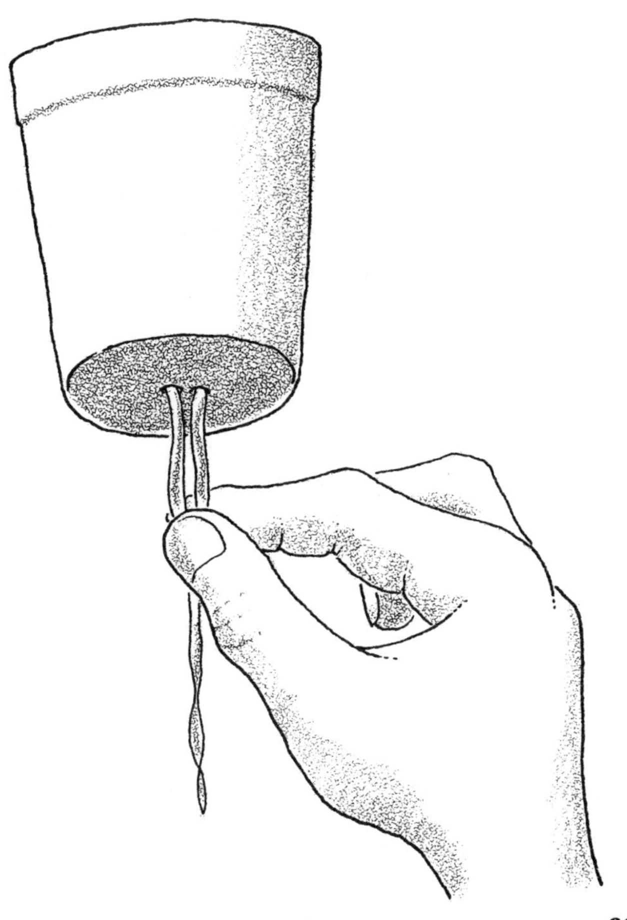

Fishing for "Clippies"

With this experiment, you can fish at your kitchen table, on your back porch, or anywhere you choose.

You need:

6 paper clips (colored ones make fishing more fun)
a large bowl of water
a facial tissue
a pencil

What to do:

Straighten one of the paper clips, shaping one end into a hook. Open and lay the facial tissue across the bowl of water.

Next, quickly but gently, place the remaining paper clips, one at a time, on top of the tissue. Then, tap around the edges of the tissue with a pencil until the paper sinks to the bottom of the bowl, leaving the "clippie fish" afloat.

Now that your homemade fishing pond is stocked, use the hook that you made to see how many clippies you can catch.

What happens:

If you are careful, you hook all five clippie fish. If not, the clippies escape to the bottom of the bowl.

Why:

Surface tension, that invisible skin that covers the water in the bowl, holds the paper-clip fish within reach of your hook. As long as you don't break that tension, you can keep fishing and could end up with the "Catch of the Day."

Unfortunately, if you break the surface tension with your hook, the weight of the clips causes them to sink.

Water-Drop Art

While you have probably only used water to clean up after an art lesson, the following experiment will show you how to create a picture using water as the main ingredient.

You need:

a large piece of waxed paper
a wooden toothpick
5 small cups of water
red, green, yellow, and blue food coloring
an eyedropper or straw
paper towels (optional)

What to do:

Put three or four drops of red coloring into one of the four cups of water and do the same with the green, yellow, and blue coloring to make four cups of "water paints." Leave the fifth cup of water clear.

Next, spread the waxed paper out on a flat surface. With the eyedropper, put three or four drops of each color water paint on the waxed paper. Remember to rinse the eyedropper in the cup of clear water first when you change colors.

Continue by dipping one end of a toothpick in the cup of clear water and then putting it *near*, but not touching, a water drop.

What happens:

The water drop moves towards the toothpick, gliding easily over the waxed paper, to help create a picture or a design.

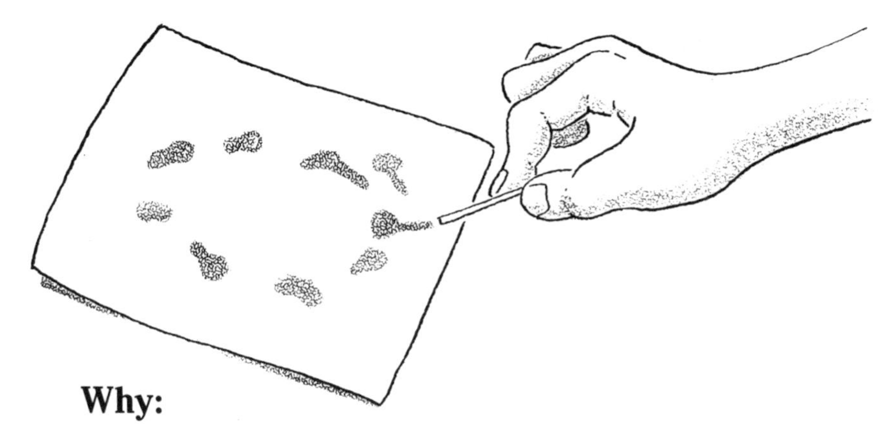

Why:

The water drop rolls around on the waxed paper because the wax keeps it from soaking in.

The water drop is drawn to the wet toothpick because of cohesion, the tendency of molecules that are alike to stick together.

What next:

Continue creating your picture or design using water drops in each of the four colors.

When you are finished, you can save the pattern by laying a paper towel over it and letting the towel absorb it. If you don't want to save your work, you can let the different-colored drops touch each other—and watch as they gobble each other up.

Oil vs. Water

"Oil and water don't mix." After this experiment, next time you hear anyone say that, you'll know why.

You need:

an eyedropper or straw
a glass of water
rubbing alcohol
cooking oil

What to do:

Pull a few drops of alcohol into the eyedropper and then watch as you release them slowly just *below* the surface of the water in the glass.

Next, refill the eye dropper with a few drops of cooking oil and add it to the water just as you did the alcohol.

What happens:

The alcohol just disappears. The oil forms drops that float to the surface and stay there.

Why:

The molecules of alcohol dissolve, fitting perfectly into the spaces between the water molecules, to form a solution.

The molecules of oil, however, do not fit together with the molecules of water, so the lighter molecules of oil are pushed into the round drops by the water and float up to the surface. There, the oil drops make little round mounds. These flatten out eventually because of the pull of gravity on them. Oil's surface tension is very weak compared to water's.

Make a Waterwheel

Huge waterwheels are used in large rivers to generate or produce a certain kind of electricity called hydroelectric power. You can make a model of a waterwheel in your sink.

You need:

a plastic-foam or plastic-coated plate
scissors a pencil a water faucet

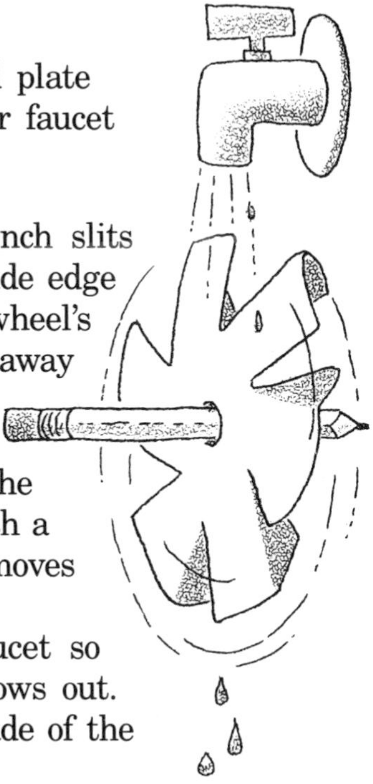

What to do:

Using the scissors, cut six 1-inch slits spaced evenly around the outside edge of the plate to form the waterwheel's blades. Bend these blades away from the plate to make them more efficient. Next, push a pencil through the center of the plate and work it back and forth a few times so that the pencil moves easily.

Now, turn on the water faucet so that a fast stream of water flows out. Hold the pencil so that one blade of the plate catches the water.

What happens:

The waterwheel will begin to spin!

Why:

The water tumbles out of the faucet, pushes against one blade of the plate, then another, and another, until the waterwheel is powered into motion. This motion can be used to generate more power. That is why electricity plants are built next to dams or fast-flowing rivers.

Deep-Bottle Diver

If you have ever played submarine in the bathtub, you will like the following experiment.

You need:

a 2-litre plastic bottle with cap
an eyedropper
tap water

What to do:

Fill the bottle to the top with water. Pull a little tap water up into the eyedropper and place it inside the bottle so that the dropper floats near the top. Replace the bottle cap.

Now, press the sides of the bottle in and then release them.

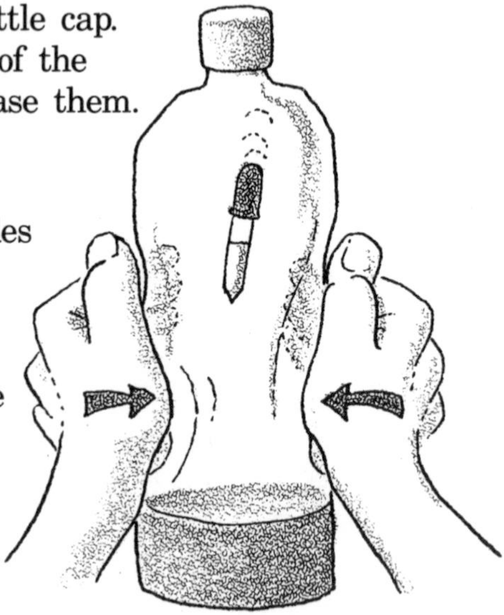

What happens:

When you press the sides of the bottle, the eyedropper "diver" sinks to the bottom.

When you release the sides, the "diver" rises again.

Why:

When the sides of the plastic bottle are pressed in, it increases the water pressure equally throughout the bottle and some additional water is forced into the eyedropper. This makes it "dive."

The amount of water in the eyedropper can be controlled by changing the water pressure in the bottle: pressing = more pressure; not pressing = less pressure.

The Warm and Cold of It

When swimming in a lake, have you ever down deep beneath the smooth surface and found the water there to be suddenly colder? Here's why.

You need:

2 small balloons
2 large, tall jars
cold and warm tap water

What to do:

Fill each of the balloons with cold water, loop the end over and tie a tight knot to keep the water inside. (If your tap water is not very cold, put some in a pitcher and add ice to cool it before filling the balloons.)

Now, fill one of the jars about halfway with warm water and the other one halfway with cold water. Place a water-filled balloon into each jar.

What happens:

The balloon filled with cold water sinks to the bottom of the jar of warm water but floats in the jar of cold water.

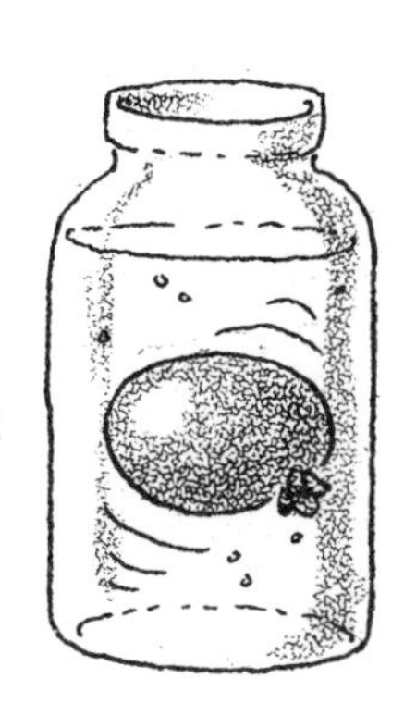

Why:

Cold water is heavier than warm water because the cold-water molecules are denser, that is, stick closer together. So the weight of the cold water in the balloon drags it down to the bottom of the jar of warm water. *Where* the balloon floats in the jar of cold water depends on the temperature difference of the water in the balloon and the jar.

Make a Purple People Eater

Sometimes, on a dull, rainy day, it's fun to mix up a Purple People Eater.

You need:

red food coloring
blue food coloring
⅔ cup of tap water
a mixing bowl
a spoon
1 cup of cornstarch
2 marbles (optional)

What to do:

Put 3 drops of red and 3 drops of blue food coloring in the water to make purple.

Pour the cornstarch into a mixing bowl. Slowly add the water, stirring to mix well.

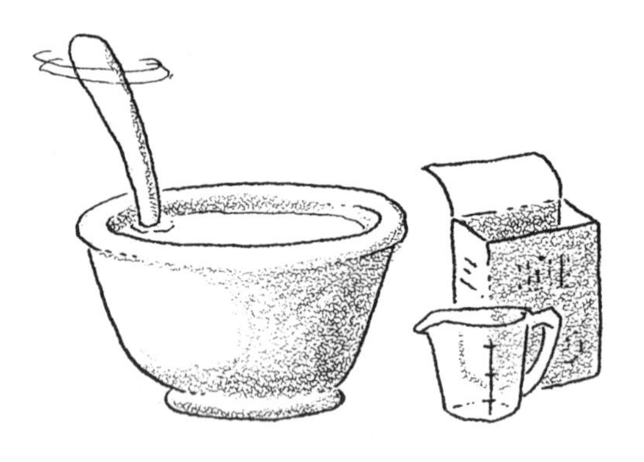

Now, grab a handful of the mixture and form a ball by rolling it between your hands. Stop rolling and let the mixture rest on your outstretched palm.

What happens:

While you are rolling the mixture between your hands, it feels dry. When you stop rolling, the ball suddenly turns into an ooze!

Why:

As you have already learned, both salt and sugar dissolve in water to form solutions. Cornstarch, however, does not form a solution with water.

Instead, the cornstarch particles are simply held together by the water, creating a mixture called a suspension. When you roll the mixture in your hands, it squeezes together on all sides and feels dry. But, when you stop rolling, the cornstarch particles in the mixture drift apart, creating an ooze.

What now:

Press two marble "eyes" into your Purple People Eater, and you will have created a monster that any Dr. Frankenstein would be proud of.

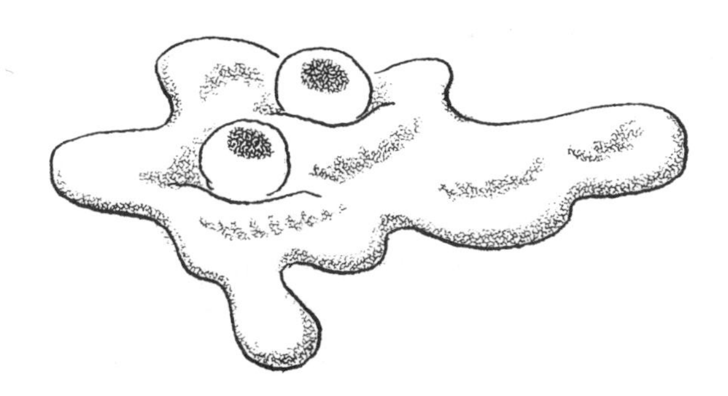

SEEING THE LIGHT

Like heat, sound, and electricity, light is a form of energy on which all life on Earth depends. Unlike sound, light can travel where there is not even any air, and it can do so at the fastest speed known—186,000 miles per second.

We see an object because light reflected from that object reaches our eyes. Over the centuries, people believed that light consisted of tiny particles. Others thought differently, although nearly everyone agreed that light travelled in straight lines. Today, scientists have learned that light sometimes acts like particles and sometimes like waves. Maybe you'll be the one to discover someday what light really is.

Sometimes Bigger Is Better

Have you ever wondered what is in some of the food you eat? Many people do, but they don't read the ingredients listed on the side of the package because the print is so small it is hard to see. The following experiment will help you to read more about it.

You need:

a food box or package

a clear glass of water

What to do:

Hold the food box up close to the glass and look through it at the ingredients list printed on the box.

What happens:

The print on the package appears much larger and is easier to read.

Why:

Because the drinking glass is curved, the light rays enter at an angle and are bent as they pass from the glass to the water. This is called refraction. The result is a homemade magnifying "glass."

Make Your Own Movie Screen

Have you ever wondered, at a movie theatre, how that narrow beam of light from the projection booth at the back could light up and fill a whole movie screen from so far away? Here's how.

You need:

a shoebox with a lid
a ruler
a pencil
a small knife or scissors
a flashlight
graph paper (lined to make squares)
a helper

What to do:

With the ruler and pencil, measure and mark out a small square in each end of the shoebox and carefully cut them out.

Replace the lid of the box. Turn on the flashlight and aim the light through the square cutouts in the box while a friend or helper on the other side of the

box "catches" the light on a piece of graph paper.

Ask your friend first to stand close to the box, then step back one ruler length from the box, then another ruler length. Watch the light on the paper.

What happens:

As your friend moves farther away from the shoebox, more and more of the squares on the graph paper are lit up, but the light's brightness begins to fade.

Why:

Once the light beam passes through the holes in the shoebox, it begins to spread out. The light loses more and more of its brightness as the same amount of light from the flashlight spreads out to fill a larger area.

Also, once it reaches the paper, part of the light's energy is absorbed by the molecules in the graph paper. Other light rays bounce, or reflect, off the graph paper and scatter around the room. This is called diffusion. It is your eyes' ability to detect this reflected light which allows you to see.

The Reappearing Penny

This is a fun trick to play on your friends because your hands never touch the penny.

You need:

a small dish (**not** see-through)
a table or countertop
a penny
tap water
a friend

What to do:

Put the penny in the empty dish and set it on a table or countertop. Next, while watching the penny, have your friend back away from the dish slowly until its edge just blocks the view of the penny.

Tell your friend not to move, and you'll make the penny magically reappear. Then, slowly fill the dish with water.

What happens:

Your friend will see the penny gradually come back into view.

Why:

When you put water in the dish, it bends the light reflected from the penny around the edge of the dish to reach your friend's eye again.

Big Bold Letters

Big, bold, or fat print is easier to see than small, skinny print. Here's how to make small letters appear fatter.

You need:

vegetable oil
a page from a magazine
(ask before you rip!)
an eyedropper or straw
tap water

What to do:

Using your finger, place a dab of vegetable oil on a word on the magazine page. Gently, rub it into the paper.

Now, using the eyedropper, put one drop of water on top of the oil-coated word.

What happens:

When you read the word through the water drop, it looks bigger.

Why:

The oil you rubbed in has coated or conditioned the paper so that the water does not soak in. As a result, the drop of water sits on top of the word and forms a lens. This lens changes the path of the light that reflects off the page and reaches your eyes and makes the word look fatter.

This same principle is used in making eyeglasses, except that it is glass, not water, that bends the light, causing it to reach our eyes at the correct angle to see better.

Amazing 3-Ring Light Show

When blue, red, and green paint are mixed together, they make black. When blue, red, and green *light* are mixed together . . . a surprising thing happens! Can you guess what?

You need:

3 flashlights, same kind
sheets of cellophane paper or
  transparent plastic wrap—
  1 blue, 1 red, and 1 green
3 rubber bands
a white wall
a small table or bench

What to do:

Using the rubber bands, fasten a different-colored sheet over the head of each flashlight.

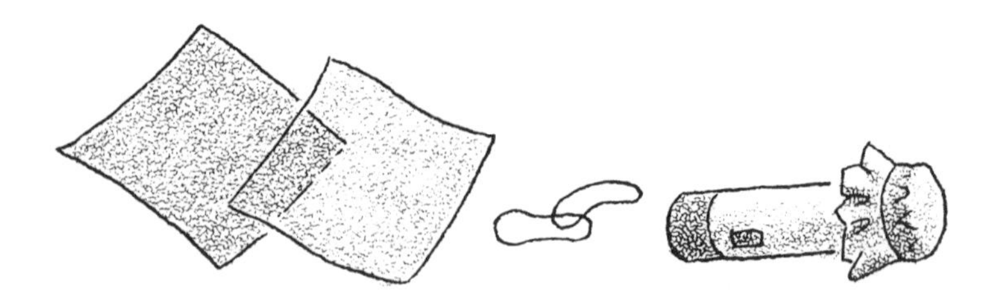

Now, place the three flashlights on a table or bench, or get one or two friends to help you, and aim the flashlights at the wall. The two outside flashlights should be turned slightly towards the middle one, which should be aimed straight ahead.

Then, turn on all three flashlights, moving the two outer ones so that the three circles of light on the wall overlap.

What happens:

A rounded triangle of white light appears in the middle of the three overlapped colored circles.

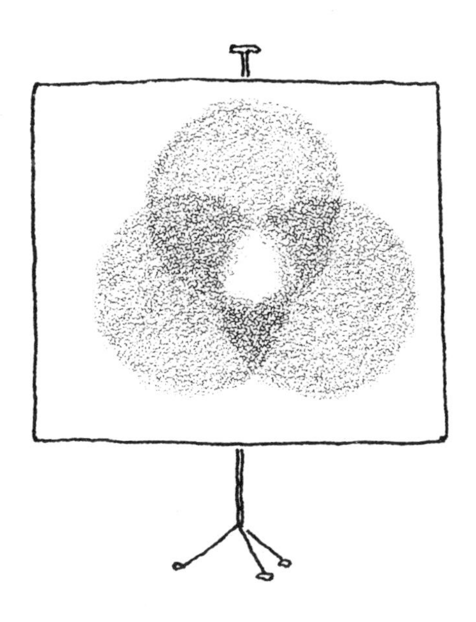

Why:

Light has within it all colors, which is why it sometimes makes something called a continuous spectrum, in other words, a rainbow.

By passing the white light through the red sheet, only red light comes out. The same thing happens with the other color sheets, so you get the three primary colors: red, green and blue. When all the colors come together again, in the middle of the lights, the mixture of all the three colors forms a rounded white triangle.

Me and My Shadow Puppets

Shadow puppets are great fun and don't require a lot of complicated equipment to put on a real show.

You need:

- cut-outs of people, animals, anything, from magazines (ask before you cut), or you can design your own on cardboard and cut them out
- small stirrers or ice-cream sticks
- glue
- a flashlight
- a wall in a darkened room
- a large book or low bench

What to do:

Glue each cut-out figure to a stirrer or stick.

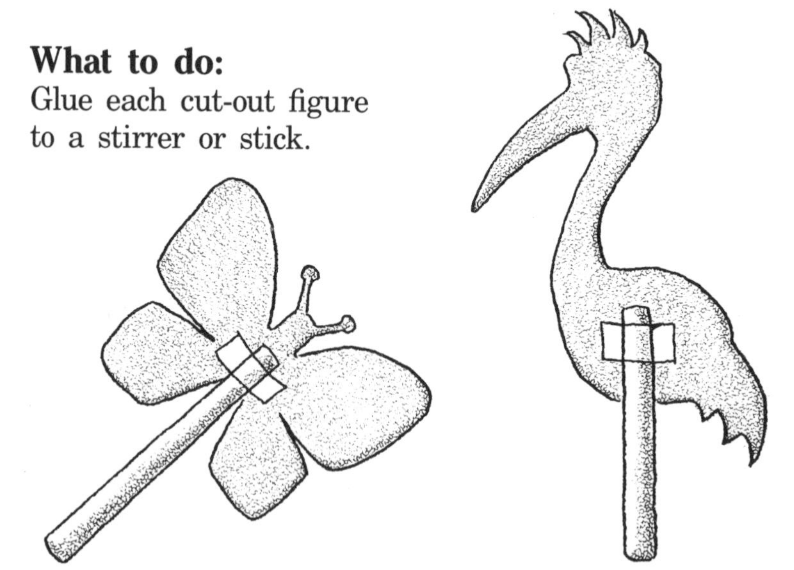

Use something like a large book or a bench to hold the flashlight. Place it three or four giant steps away from a wall. Then, turn on the flashlight and aim its beam at the wall.

Now, sit on the floor close to the wall and hold your puppets up inside the spotlight.

What happens:

Instead of being able to see your actual cut-out puppets, with colors and details, you see the outline shapes of your puppets as shadows on the wall.

Why:

Light can pass through transparent substances, such as glass or water, but other substances, such as cardboard or heavy paper, stop light from passing through. These substances are called opaque. Shadows appear behind them because light travels in straight lines and does not bend around corners.

Try holding a puppet very close to the wall and then moving it back, away from the wall. What happens to the puppet's size?

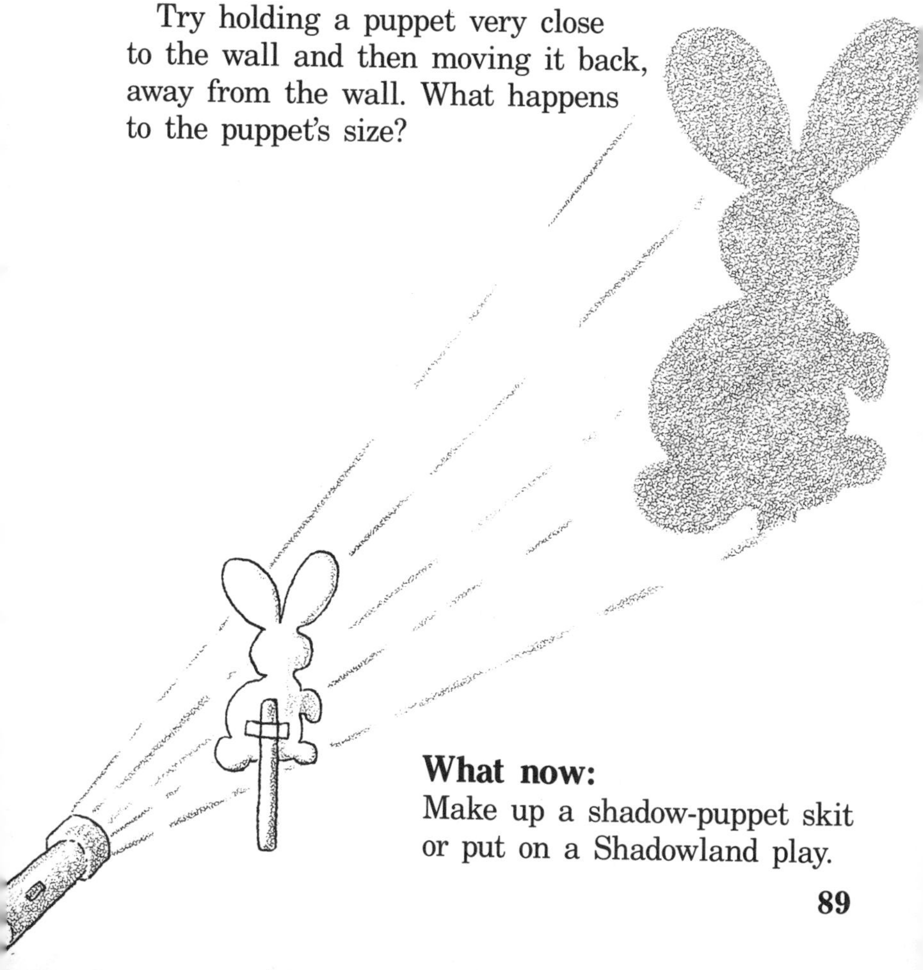

What now:

Make up a shadow-puppet skit or put on a Shadowland play.

SOUNDS LIKE FUN

What you know as sound are vibrations that travel to the ear via a sound carrier. Because there is so much air, it is the most common carrier of sound, but it is also the slowest. Sound travels four times as fast in water as it does in air. Sound travels faster at high temperatures, but slower high in the atmosphere or on mountaintops where there are fewer air molecules to vibrate.

Sound waves from vibrating objects are sent out in all directions. If you could see them, the waves might look like the circles, or ripples, that spread out when a rock is thrown into a quiet pool of water. The object making the vibrations, or sound, would be in the center of the smallest circle.

Sound is measured by decibels. These measurements go from 1 (a sound that can barely be heard) to 130 or more. A sound that measures 120 decibels hurts most people's ears. Some sounds can be so high, or squeaky, in pitch that people can't hear them, but some animals can.

Deep "C", High "C"

Sound is made up of waves that move through the air much like the rippling circles that move across the surface when a rock is thrown into a quiet pond.

You need:

a large jar

a small jar

What to do:

First, hold the opening of the large jar to your mouth and hum into it; then hum the same way into the smaller jar.

What happens:

A deeper sound is heard when you hum into the large jar, and a higher sound when you hum into the small jar.

Why:

The pitch of the sound depends on the height and diameter of the jar. Because there is more room in the large jar, your humming makes longer sound waves, so you hear a deeper, lower sound in that jar.

The sound waves in the smaller jar have less room, so they are shortened, and the frequency, or pitch, of the sound you hear is higher.

What now:

Try some other empty jars. How do they sound?

Catching Sound

The following experiment will show you how you can produce an echo, without leaving home in search of a canyon.

You need:

a helper
two paper-towel tubes
a wall
a loud "ticking" clock or kitchen timer

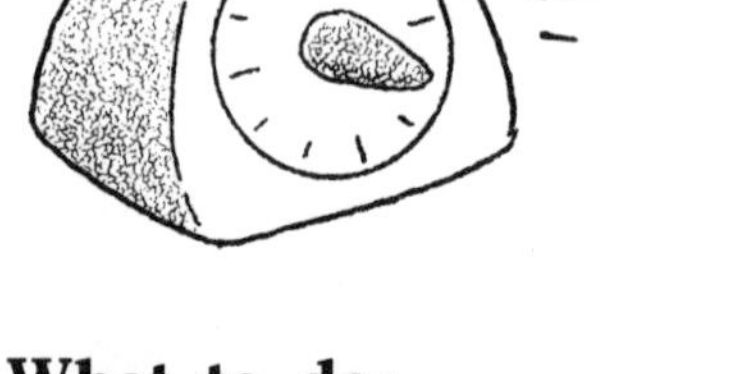

What to do:

Place one end of each cardboard tube on a slant against a wall, so that they come to a point, or make at least a 45-degree angle.

Have your helper hold the ticking clock at the other, open, end of his or her tube.

Listen at the open end of your tube.

What happens:

You can clearly hear the ticking sound of the clock through the cardboard tube.

Why:

Normally, the sound waves sent out into the air by

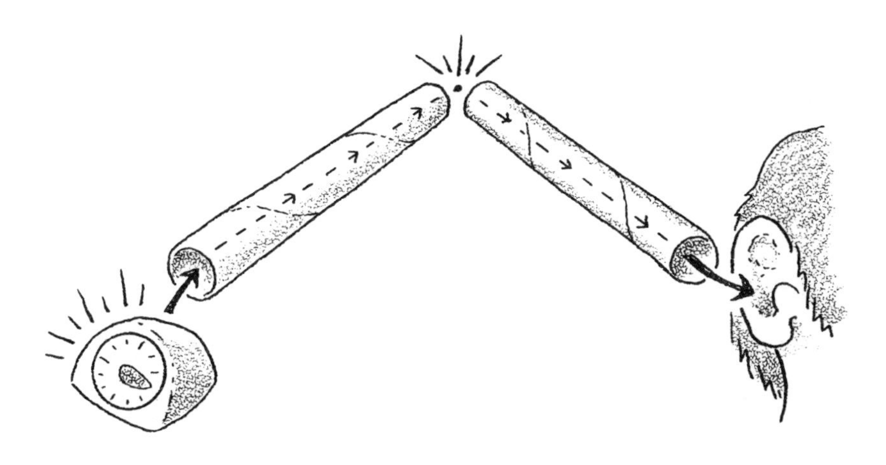

the ticking clock would scatter in all directions, becoming fainter the farther away from the clock they travelled.

By using the tubes, the sound waves of the clock's ticking are captured and directed down one tube. Then, if the tubes are held correctly, the sound waves bounce off the wall and shoot up through the second tube for you to hear—just like an echo.

Change places, now, so your helper is on the receiving end of the bouncing sound waves.

The Amazing Humm-o-comb

You don't need private lessons, sheet music, or even a teacher to learn how to play this musical instrument, and making it is a snap.

You need:

tissue wrapping paper

a pocket comb

What to do:

Wrap the tissue paper around the comb, teeth side down.

Now, hold the comb up against your lips and humm loudly.

What happens:

Even though you aren't blowing on the tissue paper at all, you feel it vibrating. Your humming sounds different, too.

Why:

The tissue paper vibrates, just like your vocal cords, because sound waves from your humming are hitting it. The vibrations of the tissue paper molecules add a whole new sound to your humming.

What now:

Don't stop with one selection. Any song that you know can be played on a comb. Get several friends together and put on an Amazing Humm-o-comb concert.

Was It Ripped or Torn?

You snag your clothing on a nail or something at school. You can tell right away, without looking, if it was ripped or torn. How? Physics tells us, listen to the sound.

You need:

an old rag or piece of fabric (ask before you use it)

What to do:

Hold the edge of the cloth in your two hands and pull evenly, tearing the cloth slowly.

Now, grab the cloth and yank it apart suddenly, so that it rips in your hands.

What happens:

When you pulled slowly, the tearing cloth made a lower sound than when you pulled suddenly.

Why:

Each time a thread in the cloth is broken, the molecules in the air around it are set into motion. When the cloth is torn, the air molecules are not bumped around as fast and the sound, or pitch, is lower.

When the cloth is ripped, however, the air molecules are bounced around at a higher rate of speed, and a higher-pitched sound is heard.

Make Your Own Sound Studio

Do you like to sing in the shower? Usually your singing sounds better there than it does in the family room. How come?

You need:

a tape recorder with a separate microphone
a new or clean metal bucket

What to do:

Hold the microphone and turn on the tape recorder. Sing any song that you like. Turn off the tape recorder.

Still holding the microphone, pick up and put the bucket over your head. Then, turn on the tape recorder again. (You might need a friend to help you here, with that bucket over your head!)

Now, sing the same song that you sang before. Remove the bucket and turn the recorder off when you are finished singing. Rewind and play through the tape.

What happens:

Your song sounds richer in tone and louder in volume when you sing with your head in the bucket.

Why:

The sound waves from your voice cause the molecules in the metal bucket and the air inside it to vibrate and "build up" your normal voice, much like a professional sound studio does for recording artists. The same thing happens when you sing in the shower.

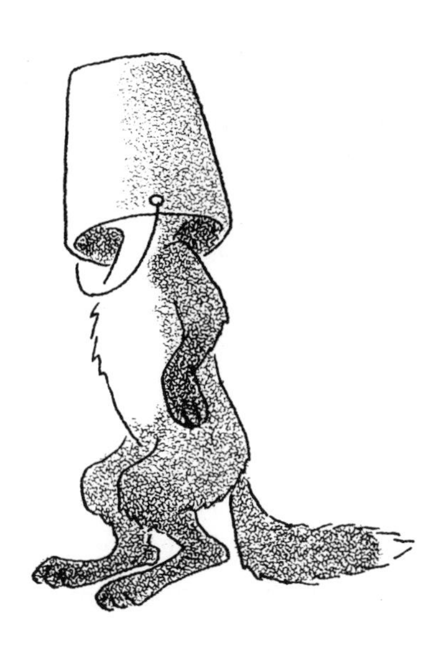

Cigar-Box Guitar

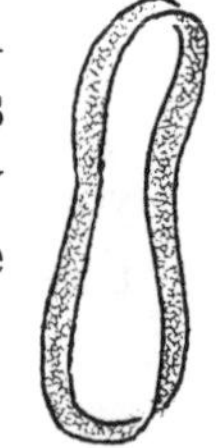

Years ago, many people made their own musical instruments. They loved music, but instruments were expensive and they had little money to buy them. Today, right now, you can do the same thing.

You need:

a cigar box (or similar box with rigid sides)
6 assorted rubber bands (including one very wide and another very narrow)

What to do:

Open the lid of the cigar box and keep it open (or remove it altogether).

Now, beginning with the widest rubber band, then the next-to-the-widest, place them lengthwise around the cigar box. Try to space the rubber bands equally, about one finger width, apart.

When you have all six "guitar strings" in place, give each of them a pluck.

What happens:

The wide rubber band has a low sound, the very narrow "string" has a high sound, and the sounds of the other rubber bands are somewhere in between.

Why:

The wide rubber band has a low vibration rate, and does not produce many sound waves.

The narrow rubber band, however, has a high vibration rate and produces a higher number of sound waves with a higher tone or pitch.

But that's not all. Pitch also depends on the degree of tightness, or tautness, of the "string." A wider but short rubber band that is pulled very tight might make a higher sound than a narrower rubber band that is put on more loosely.

What now:

Listen to the sounds of each of your instrument's strings again. If needed, switch them around so that the sounds are in order, from the lowest to the highest. When you are ready, sing a tune and accompany yourself on your cigar-box guitar.

The Silence of Snow

Is it quieter outside after a snowfall? It's true that the snow does provide a soft carpet for your feet, but what about all the other noises?

You need:

2 paper cups
18 or 20 cotton balls
white glue
a whistle

What to do:

Glue the cotton balls to the inside of one paper cup until it is completely covered with the cotton "snow."

Then, blow the whistle inside the cup with *no* cotton stuffing and listen to the sound. Next, put the whistle inside the snow-covered paper cup and blow again.

What happens:

The whistle's sound in the first cup is loud.

Inside the "snow-filled" paper cup, however, the whistle's shrill notes sound like a flute being played in a closet.

Why:

Like snowflakes, cotton balls have hundreds of tiny spaces between them. Sound waves get trapped in these tiny spaces and are muffled, like in winter when you can't hear someone trying to talk to you through a heavy scarf or muffler.

Now, you also know why there is a law requiring all cars to have mufflers.

Natural Vibrations

Do you know why soldiers deliberately walk out-of-step rather than march when crossing a bridge?
Read on.

You need:

2 small soda bottles
(same kind)
a helper

What to do:

Hold one of the small bottles to your ear and listen while your helper blows across the mouth of the second bottle until it makes a clear sound.

What happens:

The bottle that you are holding to your ear will vibrate "in sympathy," making a similar but weaker sound than your helper's bottle produced.

Why:

Depending upon its size and shape, each object has its own natural rate of vibration.

When two objects have the same vibration rate, like the two similar bottles, one can cause the other one to vibrate. When this happens, the two objects are said to be "in resonance."

The officers of soldiers approaching a bridge know that the bridge has a crumpling point related to its natural vibration rate. If the soldiers' even footsteps should happen to match the bridge's natural vibration rate, it could begin to swing and collapse! So soldiers are ordered to walk naturally as they cross the bridge.

Dance, Sprinkles, Dance

Does certain music make you want to dance like crazy? It's not wild—it's physics!

You need:

a small, round oatmeal box, with top
a sharp knife (ask permission)
candy sprinkles

What to do:

Cut an egg-size hole in the side of the oatmeal box, about two finger-digits up from the bottom. Next, put a few sprinkles on the top of the box.

Now, put your mouth to the hole in the box and hum loudly. Begin with low notes and then move on up to the higher ones.

What happens:

Somewhere, as you go up the scale of notes, the candy sprinkles will begin to dance!

Why:

The molecules on the box top are set into motion when one certain note is reached. That particular note is called the box top's resonant frequency—a note that makes it all happen. When the box top vibrates, the sprinkles dance.

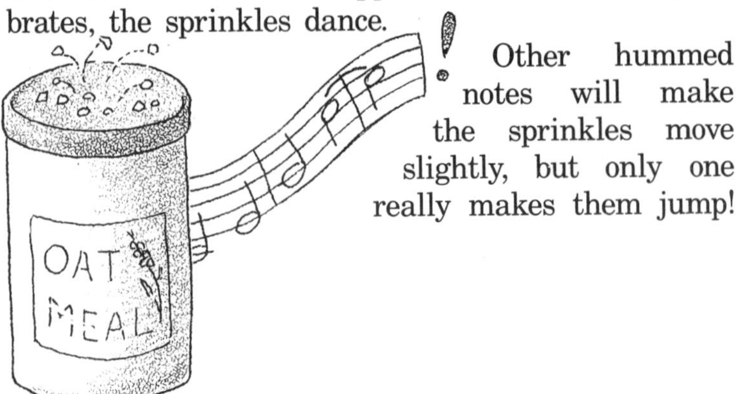

Other hummed notes will make the sprinkles move slightly, but only one really makes them jump!

Musical Nails

Visitors usually knock or ring the bell at your house door, but have no way to announce their arrival at a screened porch or patio door. A nail chime is the perfect welcoming sound.

You need:

a wire coat hanger
10 different-size nails
10 pieces of string

What to do:

Tie a string to each nail. Tie the other end of each piece of string to the straight lower bar of the coat hanger.

Hang the coat hanger on a door knob and open and close the door.

What happens:

You hear a pleasant chiming sound as the nails jingle.

Why:

When the door is opened and closed, the movement causes the nails to hit against each other and the door frame and vibrate. Each nail, depending on its size and what it is made of, makes a different note. All the nails vibrating together produces the chiming sound.

If the pieces of string were not attached to the coat hanger, and you just held them in your hand, the nails would not vibrate together or as long. Also, the sound from each nail would be much softer and quieter.

Dancing Cereal Puffs

While air molecules are usually invisible, in this experiment, disguised as cereal puffs, they dance and pass along the sound.

You need:

10 pieces of puffed cereal
10 pieces of thread
a coat hanger
a rubber band

What to do:

Tie a piece of thread around each piece of cereal.

Tie the other ends of the threads onto the straight lower bar of the coat hanger. Hook the coat hanger onto a shelf or the back of a chair, or ask someone to hold it for you.

Then, holding one end of the rubber band in your clenched teeth, pull the other end of the band towards, but not touching, one of the middle pieces of puffed cereal. Pluck the stretched band like a banjo.

What happens:

The middle cereal puff begins to move back and forth, touching its neighboring puffs on each side.

Why:

The vibrations of the plucked rubber band stir the air molecules around and inside the middle piece of cereal. Like the invisible air molecules, the puff dances from side to side and passes along the vibrations to the other nearby puffs.

The cereal pieces will continue to swing until all the vibrations, or energy, from the plucked rubber band are gone.

If the rubber band is plucked harder a second time, more of the cereal pieces will move because the sound and vibrating air molecules will travel farther. None of the cereal puffs, however, will sway very far.

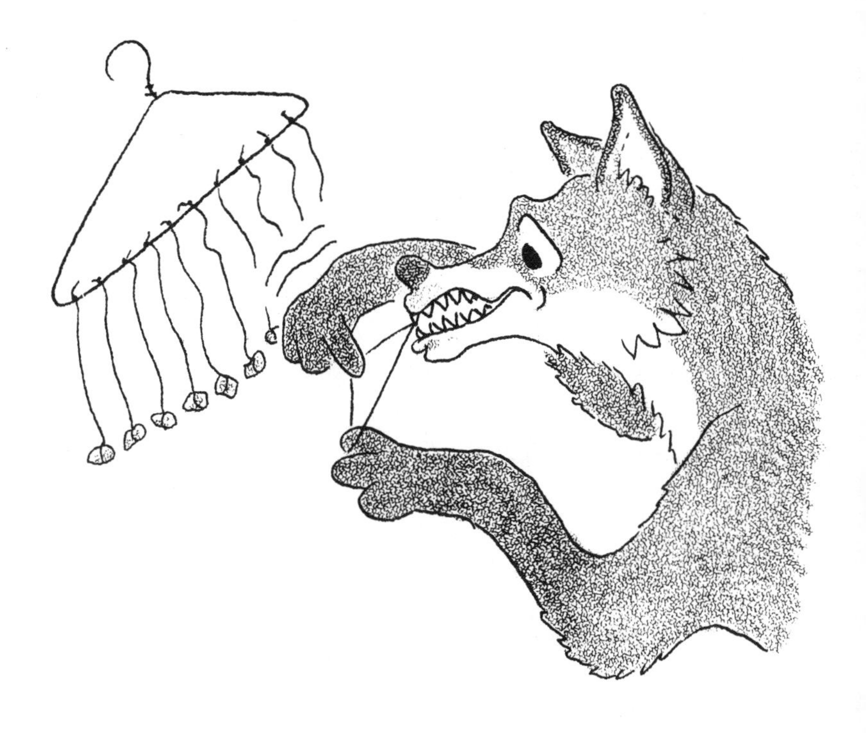

Make a Megaphone

Have you ever cupped your hands around your mouth when you wanted to yell a message to someone far away? You were on the right track.

You need:

a clean plastic milk or bleach container
a large rubber band
heavy scissors or kitchen shears

What to do:

Put the rubber band around the middle of the plastic jug to serve as your cutting guide. Then, carefully, force the sharp point of the scissors through the lower part of the jug, below the rubber band.

Followng the edge of the rubber band, cut the jug apart. Discard the bottom half of the jug.

Now, speak in a normal voice to someone across the room. Then, talk into the mouth of the megaphone that is aimed across the room.

What happens:
Your voice is louder and can be heard farther when the megaphone is used.

Why:
When you just speak, the sound waves ripple out in all directions, getting weaker the farther out they go.

The megaphone, on the other hand, aims all your sound waves in one direction, like a baseball hit to center field. Sound waves sent through a megaphone, therefore, lose less of their energy in transit and arrive with more volume.

What next:
Create an original design for yor megaphone, using markers, stickers, and cutouts. Decorate it with your school colors and your name. Don't forget to take it with you next time you go to the ballpark.

Wild Animal Calls

If you like to play jungle safari, here is a way to make your adventure sound more realistic.

You need:

a large throw-away plastic cup
a piece of heavy cotton string, about 12 inches long
a pencil or nail
a toothpick
a wet paper towel

What to do:

Use the pencil to poke a hole in the middle of the bottom of the plastic cup.

Push one end of the cotton string into and through the hole, then tie that end of the string tightly around the middle of the toothpick. Pull the string back through the cup so that the toothpick fits nicely in the bottom of the cup. (Break the ends off the toothpick, if necessary, to make it fit.)

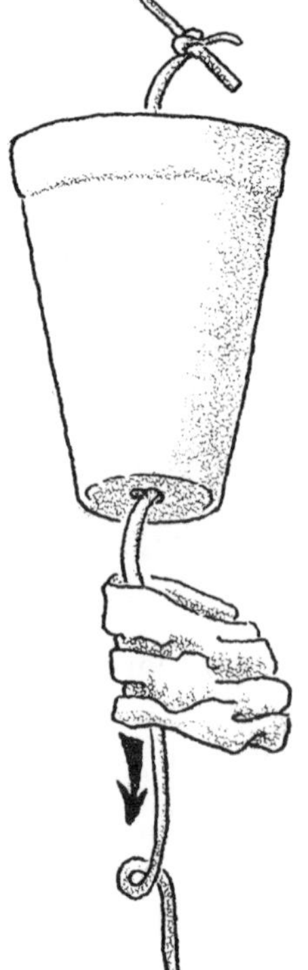

Squeeze out any excess water from the paper towel (you don't want to get the string too wet) then wrap the towel around the string near the cup.

Now, squeezing the paper towel tightly around the string, pull down.

What happens:

A loud "screaking" sound is heard.

Why:

The friction from pulling the paper towel down the string causes vibrations, which move along the string to the toothpick in the cup. From the toothpick, the vibrations travel on to the bottom and sides of the cup. Not only do these sound vibrations move farther, but they get louder because the cup also acts like a megaphone, sending the vibrations out into the surrounding air molecules.

What now:

To make a variety of "animal" calls, try using different-size plastic cups. You can also experiment with different-texture strings for some really wild sounds.

A MATTER OF GRAVITY

Gravity is the pull that objects have on other objects around them. All objects have gravity, which means that they are always trying to pull other objects towards them. The larger an object, the stronger its pull. Because the Earth itself is the largest object in our world, the pull of its gravity is the strongest we can feel.

It was in 1687 that Isaac Newton discovered and proved the existence of a force called gravity. The story goes that one day Newton was sitting under an apple tree watching the moon moving in the sky when he was almost hit on the head by a falling apple. Many people before Newton had seen apples fall to the ground, but he not only saw it happen, he also wondered why . . . and figured it out.

When you throw a ball in the air, gravity pulls it down. When you sit on the sofa, gravity holds you down, and when you walk, gravity keeps your feet on the ground. Without gravity we would all float off into outer space.

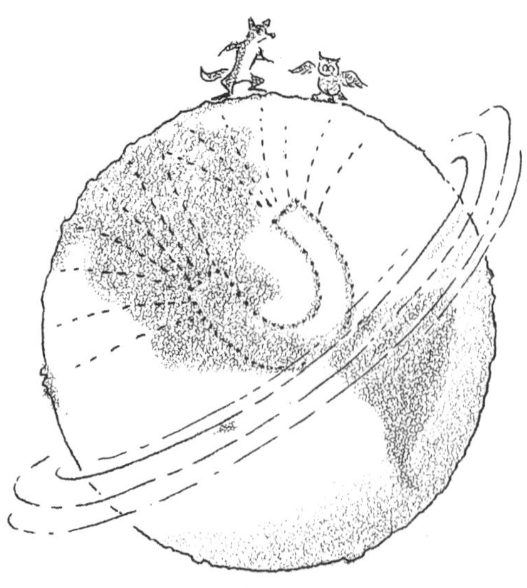

Feel the Force

You can't see the force of gravity, but it is all around you. Want to prove it?

You need:

a low, sturdy chair

What to do:

Place the low chair in front of you. Get ready, and jump up onto the chair.

Next, turn around, get ready, and jump down. Feel the difference?

Do it again. This time, when you jump down, close your eyes. Feel it this time?

What happens:

It is a lot harder to jump up onto the chair than to jump down.

Why:

Gravity is the force that pulls all objects down towards the middle of the Earth. When you jump up, you are jumping against the force of gravity, which is pulling you back.

When you jump down, the force is with you. It is doing all of the work. You really only have to step off the chair seat.

Which Drops Faster?

Aristotle, a Greek philosopher (384–322 B.C.), once believed that the heavier an object was, the faster it would fall. Was he right?

You need:

a ball of crumpled paper
a shoe
a sturdy chair

What to do:

Stand on the chair with the crumpled paper in one hand and the shoe in the other. Hold them out in front of you as high as you can and let them go at the same time.

What happens:

The heavier shoe and the lighter ball of paper both hit the floor at the same time.

Why:

Aristotle was wrong. (He was a philosopher, not a physics teacher.) An object's weight does not affect the speed at which it falls, which is a constant.

However, an object's *shape* does affect its speed. For example, if the paper had not been crumpled into a ball, the air hitting its under-surface as it fell would have slowed its rate of fall and the shoe would have hit the floor first.

Why not get another sheet of paper and try it.

Find the Center of Gravity

Want to find an object's center of gravity? It may sound hard, but it's really very simple.

You need:

any unbreakable object
a table or countertop

What to do:

Slowly push the unbreakable object to the table's edge. Keep pushing, pushing, pushing, until—

What happens:

The object suddenly falls to the floor!

Why:

When the object's center of gravity passes the table's edge, the object will fall to the floor.

Try to balance the object on the table's edge. When you do, its center of gravity has been found.

Anti-Gravity Magic

Everyone knows that water flows downwards. It has to because of gravity's pull. But here is a fun way to defy gravity and make water flow upwards . . . just this once.

You need:

2 small clear-plastic bottles
a bowl
blue food coloring (or other dark color)
an index card or piece of cardboard
access to hot and cold water faucets
a funnel
a helper

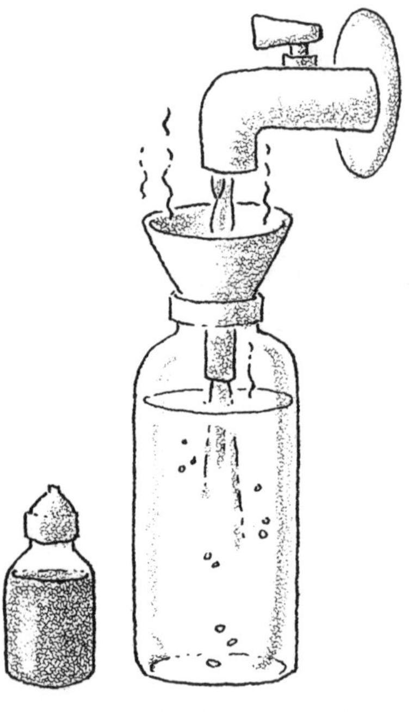

What to do:

About 30 minutes before you want to do the experiment, fill one of the bottles with cold tap water and put it in the refrigerator to make sure it's cold. When you are ready to start, turn on the hot water faucet and let the water run until it is very hot. Turn off the tap.

Put the funnel in the second small bottle, place it under the faucet, and carefully fill the bottle with hot tap water.

Set the bottle of hot water in a bowl on a table or counter top.

Add 3 or 4 drops of food coloring to the hot water. Wait a few seconds until the coloring mixes with the water.

Now, remove the first bottle of cold water from the refrigerator and place the index card or cardboard over the top of it.

Next, holding the index card *firmly* against the bottle's mouth, with your helper's assistance quickly turn the bottle upside down and set the bottle of cold water on top of the bottle of hot colored water.

Match up the two bottle tops end-to-end, then ask your helper to hold the two bottles steady while you slide the index card out from in between.

What happens:

The blue hot water *flows upward* into the cold-water bottle, seeming to defy gravity!

Why:

Hot water is not as heavy as cold water and its molecules are more active, so it rises to the top of the cold water, taking the blue food coloring with it.

Gradually, the hot and cold waters will mix until the temperature is evenly warm throughout, and the blue food coloring will be distributed between both bottles.

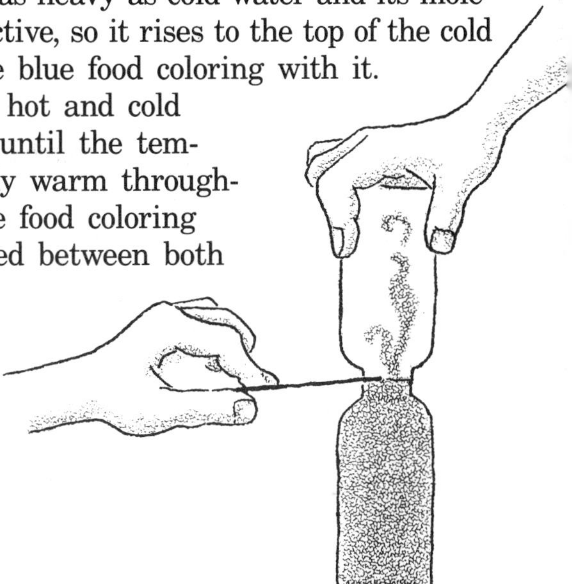

Wacky Ball

If you have ever tried to play Ping-Pong with a ball that had a nick in it, then you have played Wacky Ball.

You need:

a Ping-Pong ball
a flat-headed straight pin
a tabletop

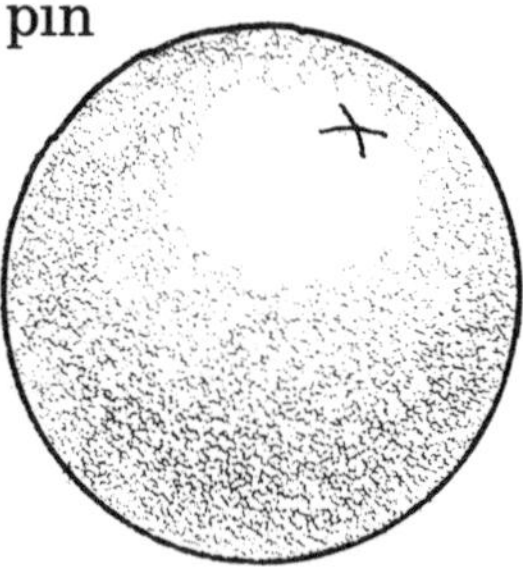

What to do:

Press the pin firmly into any spot on the ball. Now, roll the ball across the tabletop.

What happens:

The ball always stops rolling with the pin head touching the tabletop.

Why:

Before the pin was pressed into the ball, all of its weight was concentrated in its middle, the point known as the center of gravity.

With the pin pushed into the *side* of the ball, its center of gravity is shifted from the middle to the side with the pin. Now, the ball will stop rolling only when the pin is at its lowest possible point, pulled there by the force of gravity. That ball's not as wacky as it seems.

PHYSICS MIX

In earlier experiments, you have learned something about heat and air, water and light, sound and gravity. But there is still more, much more, to physics. There are laws of motion, natural rhythms, and strange forces that surround you and that you can learn to understand and use.

The experiments in this short section will just touch on these other areas of physics. Here you will discover the whirlpool in your home, the energy you create by brushing your hair, and the working pendulum. But when you finish this book, it is only a start. Physics has so much more for you to examine and study, to learn from and to marvel at. And it's all around you. All you have to do is look.

The Deep, Dark Hole

Ever try to make a hole in water? It's easy.

You need:

a mixing bowl
water
a large spoon

What to do:

Fill the mixing bowl half full of water. Then, take the spoon and stir the water quickly until it is spinning around the bowl.

What happens:

The water climbs the sides of the mixing bowl, leaving a "hole" in the middle.

Why:

When you stir, the spinning water moves away from the center of the bowl because of the outward pull of centrifugal force and forms a whirlpool, or vortex.

The "hole" that forms at the bottom of the whirlpool is smaller than the one at the top because of water pressure. The weight of the water above prevents the water below it from spreading out too much.

A Gyroscope in Your Pocket

Did you know that the reason you are able to stay up on a moving bicycle is because you are riding on two gyroscopes? Read on.

You need:

a large-size coin

What to do:

Try to balance the coin, make it stand up on end. Can you do it?

Now, hold the coin straight up and "flick" it with your finger to make it spin.

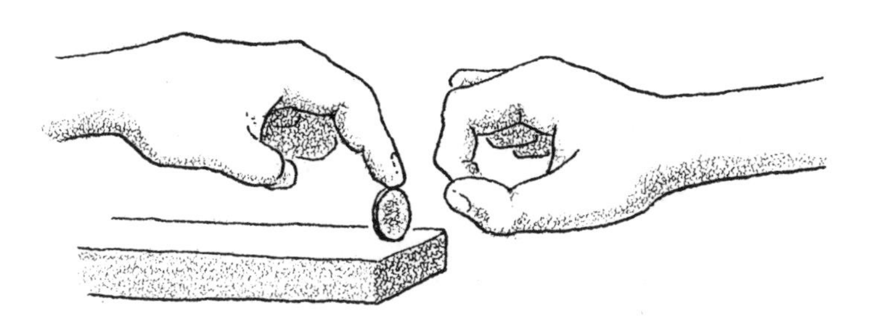

What happens:

Although the coin first fell over when you tried to balance it, the *spinning* coin balances on edge for a moment, until it slows down.

Why:

As it turns, the coin becomes a simple gyroscope, or top. The spinning motion causes the coin to stand on its end. The coin's center of gravity now runs straight down through it from edge to edge, keeping the spinning coin in place and balanced.

What now:
When you become an expert coin-spinner, try getting the coin to spin on smaller and narrower surfaces, like the end of a can or a glass, or on a book or a ruler.

Make a Balloon Rocket

A balloon rocket works the same way a real rocket does, except that it is powered by air instead of rocket fuel.

You need:
a balloon

What to do:
Blow up the balloon as full as possible and hold the end tightly closed with your fingers.

Now, let it go.

What happens:

Although it won't head for the moon, the balloon "rockets" around the room.

Why:

When you blow up the balloon, the air pressure inside presses equally in all directions and makes the balloon large and round. Everything is so perfectly balanced that the balloon simply floats as you hold it tightly.

As soon as you let go of the balloon, the air inside rushes to get out. The perfect balance of air pressure is gone.

As the air goes in one direction, the balloon goes in the other. This action-reaction motion sends the balloon zooming forward, until all of the air is gone and the balloon falls to the ground.

You have just seen Sir Isaac Newton's third law of motion, "Every action has an equal and opposite reaction," in action.

The Kissing Balloons

If kissing is frowned upon in your school, you'd better get permission for these two "kissing cousins" to visit your science class.

You need:

2 light-colored latex balloons
2 pieces of string
a piece of flannel or rayon cloth
permanent markers

What to do:

Blow up the balloons and tie a string to each one to hold them.

Now, using the markers, carefully draw a "boy" face on one balloon and a "girl" face on the other.

When the marker ink is dry, hold the balloons and rub each face several times with the flannel or rayon cloth.

Put the balloons face to face.

What happens:

The balloons will begin to kiss. They will also kiss your hair and your sweater, if you let them.

Why:

When the piece of flannel or rayon is rubbed across their faces, the friction, or rubbing, charges the balloons with static electricity. Made up of positive and negative electrical charges, which attract each other, it is very sticky stuff.

Unlike regular electricity that flows through wires and is very dangerous, static electricity stays in one place. Although it can give you a scary "zap" sometimes, it is really harmless.

What now:

Give the balloons a few more rubs, to make more static electricity, and see what other materials or things in your home the balloons will "kiss."

Want to *see* static electricity? Turn out the lights and do some experimenting in the dark.

Pendulum Sand Painting

Sand painting is an art form. Using physics, you can create your own sand painting.

You need:

- a coffee can with plastic lid
- a piece of poster board
- a broom or mop handle

a hammer
a medium-size nail
string
light sand (the clean, store-bought kind works best)
2 chairs
red sand and blue sand

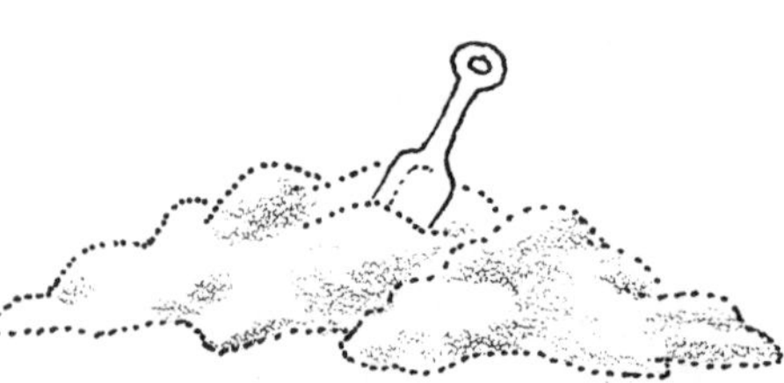

What to do:

Using the hammer and nail, punch a hole in the middle of the bottom of the can. Then, punch three holes equally spaced around the top edge of the can, just inside the rim.

Cut three short pieces of string, several inches long, and tie them to the can's rim through the holes. Gather the other ends of the strings together and knot them. Cut off a longer piece of string and tie it to this knot.

Set two chairs back to back, with a space between, and slide the broom handle through the chair's slats to hold it steady.

Then, tie the string with the can attached onto the middle of the broom handle. The bottom of the can should be just an inch or two above the floor.

Snap the plastic lid over the bottom of the can and fill the can with either red- or blue-tinted (and dry) sand.

Finally, sprinkle the poster board, giving it a thin layer of white sand, and position it underneath the coffee can. Pull the sand-filled coffee can slightly off to one side, remove the plastic lid so the sand can flow out, and release the can.

To adjust the pattern or give the can a little more "swing," give the string near the can a smooth, light push.

When one color is finished or you want to change colors, put the lid back on the can, put the second color in the can, remove the lid, and start the can swinging again.

What happens:

As the sand streams from the can, it makes a series of arcs on the poster board, and a unique sand painting appears right before your eyes.

Why:

What you have made with the coffee can is a small version of a pendulum, like Jean Bernard Léon Foucault hung in a church in 1851 to demonstrate the Earth's rotation. Allowed to swing continuously, a pendulum makes one complete circle of arcs every day.

Here, the swinging motion of the pendulum, back and forth and to the side, forms a physics-inspired pattern of ellipses on the white sand.

The Magic Water Bucket

You can pour water out, but can you sling it out?

You need:

a small bucket
some rope
water

What to do:

Fill the bucket half full of water. Then tie one end of the rope to the middle of the bucket's handle. Lift the bucket off the ground with the rope and quickly start to sling the bucket around, turning and pulling on the rope until the bucket is swinging around you about waist-high.

What happens:

The water stays in the bucket!

Why:

As you turn, centrifugal force pulls the bucket, and the water in it, upward and outward as far as the rope will allow. Because the rope holds the bucket's opening towards you, the water is pressed against the *bottom* of the bucket, even against the force of gravity. Everything is great . . . until you try to stop! As soon as you slow down, centrifugal force becomes weak or is lost and gravity takes over. That's when, if you're not careful, the water will spill.

Index